SCIENCE WITH POCKET CALCULATORS

The Wykeham Science Series

48

General Editors:

PROFESSOR SIR NEVILL MOTT, F.R.S.
Emeritus Cavendish Professor of Physics
University of Cambridge

G. R. NOAKES
Formerly Senior Physics Master
Uppingham School

Mathematics, Statistics and Computing Editor:

PROFESSOR A. C. BAJPAI
Director, CAMET,
Loughborough University of Technology

The Authors

DAVID R. GREEN is a lecturer at the Centre for Advancement
of Mathematical Education in Technology (CAMET) of
Loughborough University, UK, where he lectures to
undergraduate and postgraduate students on mathematics,
computing, and mathematical education. Previously he
was Head of Mathematics in an Essex comprehensive
school and he has also worked as a research mathematician
in the electronics industry.

JOHN LEWIS is Director of the Understanding British
Industry Resource Centre. Previously he worked as Head
of Physics at a grammar school and as a Science
Advisory Teacher for the Inner London Education
Authority, where he was responsible for developing
curricula for independent learning schemes and for
advising on the selection of electronic calculators.

SCIENCE WITH POCKET CALCULATORS

D. R. GREEN
Loughborough University of Technology

J. LEWIS
Understanding British Industry Resource Centre

 WYKEHAM PUBLICATIONS (LONDON) LTD
(A member of the Taylor & Francis Group)
LONDON and BASINGSTOKE 1978

First published 1978 by Wykeham Publications (London) Ltd.

ISBN 0 85109 560 7 (Paper)
ISBN 0 85109 660 3 (Cloth)

Printed in Great Britain by Taylor & Francis (Printers) Ltd.
Rankine Road, Basingstoke, Hants, RG24 0PR

Distribution and Representation:
UNITED KINGDOM, EUROPE AND AFRICA
Chapman & Hall Ltd. (a member of Associated Book Publishers Ltd.),
North Way, Andover, Hampshire SP10 5BE.

UNITED STATES OF AMERICA AND CANADA
Crane, Russak & Company, Inc.,
347 Madison Avenue, New York, N.Y. 10017, U.S.A.

AUSTRALIA, NEW ZEALAND AND FAR EAST
Australia and New Zealand Book Co. Pty. Ltd.,
P.O. Box 459, Brookvale, N.S.W. 2100.

JAPAN
Kinokuniya Book-Store Co. Ltd.,
17–7 Shinjuku 3 Chome, Shinjuku-ku, Tokyo 160–91, Japan.

INDIA, BANGLADESH, SRI LANKA AND BURMA
Arnold-Heinemann Publishers (India) Pvt. Ltd.,
AB-9, First Floor, Safdarjang Enclave, New Delhi-110016.

GREECE, TURKEY, THE MIDDLE EAST (EXCLUDING ISRAEL)
Anthony Rudkin, The Old School, Speen, Aylesbury,
Buckinghamshire HP17 0SL.

ALL OTHER TERRITORIES
Taylor & Francis Ltd., 10–14 Macklin Street, London WC2B 5NF.

Preface

Many people are quite content to use an electronic calculator without bothering how it works or even how it might be used more efficiently. This attitude is quite understandable for the owner of a simple four-function machine who has had little or no formal training in mathematics.

But this book is written for the owners, or intending owners, of the more advanced calculators, which incorporate some scientific and mathematical functions. The various types of calculator, together with the kinds of function they offer, are discussed and some insight is given into how the calculator actually works. After that many examples are given to show how the calculator can be used to make routine mathematics easier and to extend the scope of your work.

There now seems to be little doubt that the calculator represents a major advance in scientific and mathematical techniques. We hope that after reading this book you will agree.

A note on accuracy

We have given the results of the many calculations in this book to varying degrees of accuracy. Your calculations may be slightly different, depending on the machine being used. We would emphasize that although we may give a result to 8 or 10 significant figures we do not claim that to be necessarily entirely accurate, but rather that is what a calculator may well give. Also, we do not wish the reader to uncritically accept or quote many figures in an answer. It is a good habit to round off an answer to an *appropriate* number of figures. What is appropriate will depend upon the problem, the data and the method employed. Often it might have been more exact to have used the 'approximately equals' symbol $\approx$ where $=$ has in fact been used, but the context should indicate to the reader where this is the case. We have restricted the use of $\approx$ to cases of importance.

A note on key-stroke sequences

We have provided many illustrated calculations but your machine may behave slightly differently and key-stroke sequences may need slight modifications. Again, you may well be able to find *better* ways to perform the calculations demonstrated.

Acknowledgements

Finally we would wish to acknowledge all the help which those in the industry, our colleagues and our friends have given us. In the first group we must list John McDonald of Casio, A. G. Tamone of Hewlett Packard, Ian Jennings formerly of Texas Instruments, and others from CBM, General Instrument Microelectronics, Novus and Broughtons. Those in the last two groups are too numerous to mention individually and include staff of Loughborough University of Technology from all the following departments: Chemistry, Computer Studies, Education, Electrical and Electronic Engineering, Engineering Mathematics, Management Studies, Mathematics, Physics, Transport Technology and also a number of undergraduates who made many valuable comments on the manuscript. However we must express particular thanks to Robin Bradbeer, Jack Banks, John Lee and of course Avi Bajpai, for all the invaluable advice and encouragement which they have given us.

DAVID GREEN
JOHN LEWIS

Contents

1. The types of calculator

Originally electronic calculators were aimed at the commercial office user where they would be competing with the traditional mechanical adding machines. To compete in this market the electronic calculator had its functions limited to the four basic ones of addition, subtraction, multiplication and division, and the entry logic used was chosen so that it would be familiar to the potential operators.

As the versatility of calculators has increased their use has spread into many fields and the range of calculators available has grown, with the result that the buyer is often confused as to which one to purchase. In general, though, calculators tend to fall into six main groups, each of which is available as either desk-top or portable machines. Desk-top machines give a larger display, have larger, well spaced keys, run solely off the mains and are more expensive.

The six main groups are:

1. *Basic calculator*   Aimed at the consumer market, costing under £10 in 1978, they offer the basic operations of addition, subtraction, multiplication and division and usually percentage. Often they are provided with a 'constant function' to allow repetitive calculations, and occasionally with a memory. The display is limited to about 6 or 8 digits and the calculator is usually powered by disposable batteries.

2. *Enhanced basic calculator*   These are the first of the second-generation machines which have ousted the more limited early models. Their functions are the four basic ones with the addition of: x^2, $1/x$, a fully addressable memory, constant function and even in most cases $\sqrt{x}$, and sometimes π and brackets.

Displays tend to be 8 digit with green digitrons in preference to the original red light-emitting diodes (LEDs). Most of the major manufacturers produce models of this type at prices ranging from £7 depending on the power supply, those with disposable batteries being cheapest.

3. *Basic scientific calculator*   Realizing the potential school market —especially in science and mathematics—several manufacturers, epitomized by CBM with their 7919 machine (fig. 1.1), introduced models which offered trigonometrical functions, logarithms and

Fig. 1.1   Example of a 'semi-scientific' calculator.

powers. These must not be confused with the earlier so-called 'slide rule calculators' which were of little use because they did not offer exponential notation, also known as standard notation or scientific notation. With exponential notation it is possible to display numbers as large as $9\cdot9999 \times 10^{99}$ or as small as $1\cdot0000 \times 10^{-99}$, on an 8-digit display. Depending on the make, prices will be in the bracket £10 to £20.

4. *Advanced scientific/mathematical calculators*   Aimed at the serious student or professional user, these machines are an extension of the semi-scientific models, offering many more functions. The Texas Instruments SR 51 II, for example (fig. 1.2), has the statistical functions mean, standard deviation, factorials, hyperbolic functions and linear regression. There are also three separate memories and numerous conversions from imperial to metric units and vice versa as well as polar—rectangular coordinate conversion. A comparable calculator, the CBM 5190, is available with the added facilities of being able to deal with Gaussian (or normal), Poisson and binomial distributions, numerical integration and even complex numbers.

Fig. 1.2 The Texas Instruments SR 51 II.

5. *Specialist calculators* The first specialist calculators were aimed at the financial experts, allowing them to quickly work out such things as discounted cash flows, interest and loan rates plus all the other things which are so important in commerce. It is likely that we shall see more of this type of calculator in the future, mainly because of the ways in which the workings of a calculator are organized (which is explained more fully in Chapter 3). Since the available functions depend only on the preprogrammed Read-Only-Memory (ROM) which controls the workings of the arithmetic unit, it is possible to switch ROMs around to alter functions.

6. *Programmable calculators* This class is dealt with in more detail in Chapter 16. They do allow one to do repetitive complicated calculations without the necessity of performing all the key-strokes every time since these are remembered by the machine.

It is useful to subdivide this particular type into those which can retain the programs in some way even when switched off, and those which cannot. In the former category are the very powerful HP 67, the Texas Instruments TI 59 (which replaces the SR 52) and the Novus 7100.

These are computer-like in the programming facilities offered. The programs may be stored either on small magnetic cards which are 'read' by the machine when required or alternatively in special semiconductor memories contained in a tiny cartridge which is inserted into the calculator. When not in use the cards or cartridges can form part of an extensive software library.

In the second category (those which cannot retain the program when the power is turned off) are machines such as the Texas Instruments TI 58 (which replaces the SR 56). A relatively new development by Casio and Hewlett-Packard enables the program to be stored in the calculator's *internal* memories even though the power has been switched off. Casio do this by providing separate silver oxide batteries for use with the program storage chips only, which means that 256 program steps can be stored for up to a year. However once a new program is written, the old one is lost and can only be re-entered by going through all the key-strokes again. In point of fact, the large amount of program storage often makes it possible, by careful management, to store more than one program.

Fig. 1.3   Texas Instruments PC 1000 printer cradle with an SR 52 in place.

DISPLAYS

Manufacturers are now offering calculators in most categories which give a print-out in addition to the more usual illuminated display. With a printer—usually a thermal one where a matrix of small heaters is arranged in a similar fashion to the individual diodes of the LED display—a permanent 'hard copy' of calculations can be obtained. A major advantage in having a print-out is that the user can check through the working.

Some of the newer scientific machines from Texas Instruments are capable of being plugged into a special cradle which contains a printer (fig. 1.3). Thus a pocket calculator can double as a desk-top model with print-out, enabling it to be used in the home, office or laboratory, as well as in the field.

Most displays still employ illuminated digits. The three types of display commonly used are the light emitting diode (LED), the liquid crystal, and the green digitron.

The LED displays are made by a process which is very similar to that used to produce the calculator chips themselves and are completely solid-state, being based on gallium arsenide. The display is invariably red—although it is now possible to produce other colours—and is highly magnified by cylindrical lenses, which can limit peripheral viewing. The digits are formed by illuminating the relevant bars of a seven segment matrix (fig. 1.4).

Fig. 1.4   Seven-segment displays.

The bars may be a single solid block of semiconductor or each bar may be composed of a number of individual diodes.

Liquid crystals are a very promising display since they do not generate light and hence draw very little power. Rather they reflect or transmit incident light (so you cannot use them in the dark). In the 'field effect' type of liquid crystal cell, a nematic liquid crystal is sandwiched between two polarizing plates whose optical axes are at right angles. When the display is excited by an electric field the orientation of the liquid crystal molecules changes, effectively rotating the plane of polarization, thus allowing light to pass through or to be reflected from the mirror-like backing plate. By selectively altering the crystal in this way and using a seven-segment matrix, figures can be produced which are either light against a dark background, or else dark against a bright background, this contrast making the device visible over a wide range of lighting intensities, unlike the LED which cannot be seen in bright sunlight. The life of a liquid crystal display is limited, according to some estimates, to about five years.

A digitron is an assembly of triode valves in a glass envelope. The phosphor-coated anodes form the seven segments of the visible part of the display. A positive potential difference between the anode and cathode accelerates electrons onto the anode causing the phosphor to glow. The controlling grid and the filament wires are in front of the anode and are normally invisible. Their life, too, is reportedly limited. Both liquid crystals and digitrons give larger digits than LEDs and so may be viewed without prior magnification.

None of these displays presents a steady picture. A technique known as multiplexing is used to light each digit for a fraction of the total cycle but the flicker is so rapid that it is not noticeable.

Multiplexing can be used in this way because the data within the calculator is read in *serial* form, that is it comes one digit after another rather than in *parallel* form where all the digits are produced, and need to be displayed, simultaneously.

POWER SUPPLIES

The power supply for a calculator can be quite expensive, especially if disposable batteries are used. These batteries are available either in the traditional Leclanché cell form or as the mercury cell equivalent. The latter are more expensive but offer a longer life. None of these disposable cells should ever be recharged since an explosive build up of gas within the cell may result.

To reduce the running costs, manufacturers often provide a 'mains unit' so that the calculator may be operated directly from the mains. This is not a charger in the strict meaning of the word, since it does not charge up the batteries. The insertion of the jack plug into the calculator disconnects the internal batteries: if this is not so then the disposable batteries must be removed before using the mains unit. One manufacturer prints a warning to this effect on the disposable batteries.

Alternatively, rechargeable nickel–cadmium cells may be fitted and a mains charger unit supplied. Initially more expensive, this combination is cheaper in the long run. Note that if rechargeable cells are fitted they should be in place when the calculator is being run from the mains unit as they often help smooth the rectified voltage.

One or two machines based on disposable batteries have an automatic display turn-off that operates after some 15 sec. The display is recalled simply by pressing one of the function keys, but some people find the disappearance of the displayed number after such a short time interval rather annoying.

It is not always possible to tell that the batteries are low, because the display does not necessarily get dimmer. The first indication is often an erratic display followed by incorrect calculations. One or two machines do provide a few minutes' warning, e.g. all the decimal points light up. Even so this will normally only let you finish the current calculation. In critical situations such as examinations it is always best to ensure that either spare disposable batteries or another battery pack are available.

The mains unit should conform to British Standards in the U.K. and should have a separate mains lead. Some Continental European specifications allow the mains plug to be integral with the charger. In use, the unit may be warm to the touch although this is no cause for alarm. Always check that, if a changeover switch is fitted, it is set to the correct a.c. voltage.

In no circumstances should a calculator ever be operated from a charger other than that recommended for it. Different models from the same manufacturer may require different chargers or mains units.

KEYBOARDS

The keyboard is the interface between the user and the calculator and a lot of design work has gone into producing a suitable ergonomic layout. Originally large reed switches were used, operated by small magnets which were pushed down by the key. Today very thin keyboards have been developed which can cram fifty or more keys onto a pocket calculator. The requirement is for the key depression to have a positive feel so that the user can tell when a digit has been entered. One way of achieving this is to use the 'oil can' or 'clicker' property of a piece of curved metal so that when the key is depressed the metal snaps down to make contact. An alternative is to use elastomer keyboards and yet another way is to use a small coiled spring under each key to overcome the natural springiness of the metal contact. Whichever method is used, the contact is never a clean 'make', it bounces for a few times and this key bounce has to be allowed for electronically, otherwise spurious entries would be made.

NEW DESIGNS

Prices seem to be stable. Bargains are likely therefore to be the ends of ranges or 'new' machines built around an obsolete chip. In 1978 the life of a calculator range to a manufacturer is a year to eighteen months, after which a new design takes its place having more features and functions. Fig. 1.5 shows a possible development for the future.

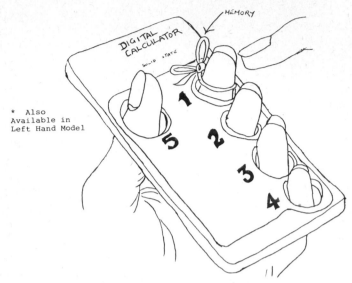

Fig. 1.5 A rough drawing of the prototype of the next generation of calculators.

2. Calculator features

The various features found on a calculator depend on which of the six categories it falls into (see pp. 1–4). Most buyers choose a calculator with too many functions, even for advanced work, and find that some of the functions are superfluous—but it is nice to know that they are available!

TYPES OF LOGIC

One major decision in the purchase of a scientific/mathematical calculator is the method of entry—otherwise known as the *language* or the *logic*. Originally arithmetic logic was available, identified by there being two $\boxed{=}$ keys on the calculator. One was $\boxed{\begin{matrix} + \\ = \end{matrix}}$ and the other, usually in red, was $\boxed{\begin{matrix} - \\ = \end{matrix}}$. This system has now become almost extinct, leaving algebraic logic and Reverse Polish logic. Both these forms have their advantages and disadvantages though the balance is slightly towards algebraic logic because it appears more natural.

It was the Polish logician Jan Lukasiewicz who first demonstrated, in 1951, how expressions could be specified by placing operators immediately *before* their operands. As an example take the expression

$$(a+b) \times (c-d)$$

This can be re-written as

$$\times + ab - cd$$

and is as unambiguous in the second form as in the first, mathematically that is. It would be read as 'multiply the sum of a and b by the difference of c and d'. Alternatively the operators may come *after* the operands in this form

$$ab + cd - \times$$

which also has the same meaning.

In honour of Lukasiewicz, the two notations have become known as Polish and Reverse Polish respectively.

Computer compiler writers have been quick to realize the advantages of using Reverse Polish logic since it means that as an expression is scanned from left to right every operator can be executed immediately it is encountered. It does not have to be stored as in algebraic logic where the operator is sandwiched between the operands. Thus the Reverse Polish Notation has advantages in needing less storage space on a calculator chip. Hewlett Packard were the first manufacturers to produce a scientific machine using Reverse Polish and they have stuck to it. Other manufacturers have gradually swung over to algebraic logic, which corresponds more closely to the written format.

To show the differences let us try some problems using both methods:

$$a+b \times c$$

For algebraic logic the key strokes:

$$\boxed{a}\ \boxed{+}\ \boxed{b}\ \boxed{\times}\ \boxed{c}\ \boxed{=}$$

will give the incorrect answer (unless you are using a Texas Instruments machine) because the expression to be evaluated really means

$$a+(b \times c)$$

in mathematical terms but the key strokes described would work out $(a+b) \times c$. So the key strokes would need to be:

$$\boxed{b}\ \boxed{\times}\ \boxed{c}\ \boxed{+}\ \boxed{a}\ \boxed{=}$$

which is inconvenient for the user, who has to rearrange the expression. Using a Reverse Polish machine, however, the key strokes are quite simply:

$$\boxed{a}\ \boxed{enter}\ \boxed{b}\ \boxed{enter}\ \boxed{c}\ \boxed{\times}\ \boxed{+}$$

but some thought is still required of the user.

Another example might be

$$(a \times b)+(c \times d)$$

and this becomes much more difficult on the algebraic machine since, if it does not have memories or parentheses, the expression has to be rearranged to become

$$\left(\frac{a \times b}{d}+c\right) \times d$$

for which the key strokes are:

$$\boxed{a}\ \boxed{\times}\ \boxed{b}\ \boxed{\div}\ \boxed{d}\ \boxed{+}\ \boxed{c}\ \boxed{\times}\ \boxed{d}\ \boxed{=}$$

or, if a memory is available:

$$\boxed{a}\ \boxed{\times}\ \boxed{b}\ \boxed{=}\ \boxed{STO}\ \boxed{c}\ \boxed{\times}\ \boxed{d}\ \boxed{+}\ \boxed{RCL}\ \boxed{=}$$

A point to note is that on modern machines there should be no need to clear before entering $\boxed{c}$.

With Reverse Polish, however, the key sequence is:

$\boxed{a}$ $\boxed{enter}$ $\boxed{b}$ $\boxed{\times}$ $\boxed{c}$ $\boxed{enter}$ $\boxed{d}$ $\boxed{\times}$ $\boxed{+}$

Many algebraic logic calculators now offer parentheses (or brackets), either singly or multiply nested, which act as a form of memory, storing operands and pending operations. The TI 58 for example can accommodate eleven operands with ten operations pending. This means that expressions are evaluated properly before being combined with other expressions.

In the example given above the key strokes on a machine with parentheses would be

$\boxed{(}$ $\boxed{a}$ $\boxed{\times}$ $\boxed{b}$ $\boxed{)}$ $\boxed{+}$ $\boxed{(}$ $\boxed{c}$ $\boxed{\times}$ $\boxed{d}$ $\boxed{)}$ $\boxed{=}$

The intermediate results, i.e. the expressions enclosed by brackets, are thus stored until required, the final combination being initiated by the $\boxed{=}$ key. If the example had been keyed in as

$\boxed{(}$ $\boxed{(}$ $\boxed{a}$ $\boxed{\times}$ $\boxed{b}$ $\boxed{)}$ $\boxed{+}$ $\boxed{(}$ $\boxed{c}$ $\boxed{\times}$ $\boxed{d}$ $\boxed{)}$ $\boxed{)}$

the final bracket would have completed the combination and there would have been no need to depress the $\boxed{=}$ key.

The provision of multiple brackets allows more complex expressions to be evaluated, such as:

$$(2 \times (1-4) + 3 \times (2+3)) - ((3 \times 2) - (4 \times 5))$$

and that only needs two sets of nested brackets.

Most algebraic logic calculators deal with the last two numbers entered—one of which may be the result of a previous operation—and combine them according to the operator keyed in before the last number entry. They make no allowance for the relative mathematical weightings of different operators. Most people are aware that multiplication and division should take precedence over addition and subtraction but the calculator is not. Except, that is, for the Texas Instruments scientific and programmable range, which have their 'algebraic operating system'. What this means is that these machines are programmed to do multiplication and division before addition and subtraction. An extra storage register is needed to service this extra facility and the hierarchy itself can cause some problems. Its best use is with the 'sum of products' example:

$$a \times b + c \times d \qquad \textit{Sum of products}$$

We have already seen how this type of problem can cause difficulty for the usual algebraic logic calculator, but with the Texas Instruments system we simply use the key strokes

$$\boxed{a} \; \boxed{\times} \; \boxed{b} \; \boxed{+} \; \boxed{c} \; \boxed{\times} \; \boxed{d} \; \boxed{=}$$

and this gives the correct result. Should one require the product of sums then things are not quite so straightforward:

$$(a+b) \times (c+d) \qquad\qquad \textit{Product of sums}$$

Keying in $\boxed{a} \; \boxed{+} \; \boxed{b} \; \boxed{\times} \; \boxed{c} \; \boxed{+} \; \boxed{d} \; \boxed{=}$, the answer obtained will be that resulting from the expression $((a+b) \times c)+d$ on ordinary algebraic machines, or $a+(b \times c)+d$ on Texas Instruments 'sum of products' logic machines, so one has to remember that a special technique is needed. The key strokes to evaluate the product of sums expression could be:

$$\boxed{a} \; \boxed{+} \; \boxed{b} \; \boxed{=} \; \boxed{\text{STO}} \; \boxed{c} \; \boxed{+} \; \boxed{d} \; \boxed{=} \; \boxed{\times} \; \boxed{\text{RCL}} \; \boxed{=}$$

(The second $\boxed{=}$ is only essential on 'sum of products' logic machines). It is this problem which has led Texas Instruments to put parentheses on their later models and these are that much easier to use, having the advantages of the 'sum of products' logic without the drawbacks. Actually the product of sums *can* be evaluated without recourse to memory or brackets—a challenge for the interested reader. (See *Solutions and Notes* at the end of the book for the details.) When functions are to be evaluated on an algebraic logic machine, the entry used is in fact in Reverse Polish form in that the number is keyed in first followed by the operator. For example, to work out sin 45° the key strokes would be

$$\boxed{4} \; \boxed{5} \; \boxed{\sin}$$

This is an important consideration since the function keys of the scientific and mathematical functions should only operate on the single number shown in the display. This will avoid combination of the displayed number with any other number or result being held in the calculator or loss of previous steps. Thus the hierarchy of the machine is extended so that certain scientific and mathematical functions take precedence over multiplication and division. Unfortunately even some quite new and otherwise excellent models do not meet this requirement with functions such as sin or log. Suppose we wish to work out the answer to 5×6^3. The key strokes one would expect to use would be:

$$\boxed{5} \; \boxed{\times} \; \boxed{6} \; \boxed{y^x} \; \boxed{3} \; \boxed{=}$$

Do this on a CBM 4190 or many other machines and the correct answer of 1080 is obtained. However, if the identical key sequence is carried out on a Texas Instruments SR 51 or CBM 7919 the answer obtained is 27000.

The SR 51 and CBM 7919 in fact solve the problem $(5 \times 6)^3$ because the $\boxed{y^x}$ key does not work solely on the displayed number, it completes any pending $\boxed{\times}$ or $\boxed{\div}$ operation first. This is because the $\boxed{y^x}$ key rates the same as the $\boxed{\times}$ or $\boxed{\div}$ key in the hierarchy of these machines.

It is in details like these that the differences between makes and models begin to show up and can cause problems if you are used to working with a particular machine and then change to a different one.

To help resolve doubts as to which entry logic—algebraic or Reverse Polish—should be used, the authors offer the following advice:

For use up to and including sixth form level	Algebraic logic.
For non-specialized use beyond sixth form level	Algebraic logic.
For specialized work beyond sixth form level	Either algebraic or Reverse Polish logic depending on personal preference.

CALCULATOR FUNCTIONS AND WHAT THEY DO

Let us now look at the functions which are currently offered on various calculators and define the tasks which they perform.

$\boxed{\text{ON/OFF}}$ As it suggests this switch turns the calculator on or off. An automatic clear is incorporated so that at turn-on all registers contain 0. One or two manufacturers provide a three-position switch to select either internal battery power or power from a mains unit.

$\boxed{\text{C}}$ *Clear all.* Touching this key ensures that all the working registers are cleared of the current calculation. Numbers stored in memories are not affected. It should always be used before commencing a new calculation.

$\boxed{\text{CE}}$ or $\boxed{\text{CLX}}$ *Clear last entry.* This enables the user to correct any mistake made during entry. It only operates on the number shown in the display. Quite a few machines combine this key with the *Clear all* key to give a dual function key. The first touch clears the last entry whilst an immediate second depression clears all.

$\boxed{+}$ $\boxed{-}$
$\boxed{\times}$ $\boxed{\div}$ The four basic *arithmetic operations.*

$\boxed{=}$ *Equals* key. Used in algebraic logic machines to complete a previously entered operation and thus to display the result.

| enter | Only found on calculators using Reverse Polish Notation. It is used to *enter* the displayed number into the stack for subsequent use. |

| % | Used to calculate *percentages*. Useful for VAT calculations, error determinations, etc. Usually used in conjunction with one of the basic operations keys. |

| √x | *Square root.* Use of this key means that the calculator will work out and display the square root of the displayed number. |

| x² | *Square.* The displayed number is squared and the answer displayed. |

| 1/x | *Reciprocal.* Calculates and displays the reciprocal of the displayed number. |

| x⇌y | *x/y interchange.* Changes the order within the calculator so that the number entered first, or a previous result, y, is interchanged with that entered second, x, and currently showing in the display. A simple example would be the key strokes |

$$\boxed{4} \quad \boxed{\div} \quad \boxed{5} \quad \boxed{x{\rightleftharpoons}y} \quad \boxed{=} \quad \text{display } 1{\cdot}25,$$

i.e. $5 \div 4$ is calculated.

Another use is to quickly calculate problems of the type $a/(x+y+z)$ in that the denominator, $(x+y+z)$, can be worked out before the numerator is entered.

The same key may also be used in some statistical and conversion sequences for data entry.

| +/− or CHS | *Change sign.* Used to allow the sign of the displayed number to be altered from positive to negative or vice versa. Also used to give the appropriate sign to an exponent if required. |

| π | *Pi.* Recalls to the display the value of this constant to the accuracy of the machine. (As a check $\pi = 3{\cdot}14159265359$ to 12 significant figures—abbreviated to 12 s.f.). |

| STO | *Store.* Stores the displayed number in the memory, replacing any number already stored there. When commencing a new calculation always use the STO key for the first entry into a memory rather than M+ if both are provided. |

$\boxed{\Sigma}$ or $\boxed{M+}$	*Memory addition* key, which, when depressed, will add the displayed number to that number already stored in the memory, the result being left in the memory.
$\boxed{M-}$	*Memory subtraction* key, which subtracts the displayed number from the number stored in the memory, the result being left in the memory.
$\boxed{M\times}$ or $\boxed{PROD}$	*Memory multiplication* key. The number already stored in the memory is multiplied by the number shown in the display, the result being left in the memory.
$\boxed{M\div}$	*Memory division* key. The number already stored in the memory is divided by the number shown in the display, the result being left in the memory. Note that with these memory keys the displayed number is retained in the working registers for use if required after the appropriate memory key has been pressed.
$\boxed{RCL}$ or $\boxed{MR}$	*Recall to the display* the number stored in the memory for use in a calculation. The memory retains the number for further use.
$\boxed{X\rightleftharpoons M}$ or $\boxed{EXC}$	*Exchange* the displayed number with that in the memory. A straight swap.
$\boxed{CM}$	*Clear memory.* If there is more than one memory then this key depression, like those of the other memory functions, will be followed by a digit to address the particular memory.
$\boxed{R\downarrow}$ and $\boxed{R\uparrow}$	*Roll stack.* Only found on Reverse Polish machines and allow the user to review the contents of the working registers or to reposition numbers.
$\boxed{(}\ \boxed{)}$	*Parentheses* or *brackets.* Used in calculations just as one would use them in ordinary algebraic calculations.
$\boxed{EE}$ or $\boxed{EEX}$	*Enter exponent.* To cope with very large or very small numbers the display mode of the calculator can be changed to exponential notation. In this there is a mantissa followed by an exponent, e.g.

$$1 \cdot 2345678 \ 55, \text{ which represents } 1 \cdot 2345678 \times 10^{55}.$$

After entering the mantissa the $\boxed{EE}$ key is depressed to allow entry of the exponent. For a negative exponent the *Change sign* key will also have to be depressed after the exponent has been entered.

$\boxed{\text{EE}\downarrow}$ $\boxed{\text{EE}\uparrow}$ *Change the exponent value* by shifting the decimal place of the mantissa. For example if the display were

$$1{\cdot}23456 \quad 8$$

by pressing $\boxed{\text{EE}\downarrow}$ once the display would become

$$12{\cdot}3456 \quad 7$$

by pressing $\boxed{\text{EE}\uparrow}$ once the display would become

$${\cdot}123456 \quad 9$$

Some machines now offer 'engineering notation' which means that the exponent is always a multiple of 3.

$\boxed{\text{sin}}$ $\boxed{\text{cos}}$ The *trigonometric functions* which should be valid for
$\boxed{\text{tan}}$ all entries of angles in all four quadrants. The simplest scientific calculators may restrict the angles to 0°–90° but better machines not only allow 0°–360° but also very much larger angles, positive and negative.

$\boxed{\text{sin}^{-1}}$ $\boxed{\text{cos}^{-1}}$ *Inverse trigonometric functions.*
$\boxed{\text{tan}^{-1}}$

$\boxed{\text{INV}}$ or   May be used to obtain *inverse function* values of the
$\boxed{\text{F}^{-1}}$ or   displayed quantity by depressing this key before the
$\boxed{\text{arc}}$ appropriate trigonometric or hyperbolic key, e.g. $\boxed{\text{arc}}$ $\boxed{\text{sin}}$ is equivalent to $\boxed{\text{sin}^{-1}}$. (On some machines $\boxed{\text{INV}}$ $\boxed{\text{ln}}$ is used for $\boxed{\text{e}^\text{x}}$ and so on.)

$\boxed{\text{log}}$ Calculates and displays the *logarithm* to base ten of the displayed number.

$\boxed{10^\text{x}}$ Calculates and displays the *antilogarithm* to base ten of the displayed number, i.e. 10 raised to the appropriate power.

$\boxed{\text{ln}}$ Calculates and displays the *natural logarithm* of the displayed number.

$\boxed{\text{e}^\text{x}}$ Calculates and displays the natural antilogarithm of the displayed number. The function is better known in its own right as the *exponential function* (e = 2·71828 . . . .).

$\boxed{\text{2nd}}$ or $\boxed{\text{F}}$ *Shift* key allowing keys to have a dual purpose. The key is depressed to access the second, or 'upper', function of a key.

$\boxed{y^x}$ Used to raise the first number entered (y) to a *power* determined by the second number entered (x). Thus the key-stroke sequence

$$\boxed{8}\ \boxed{y^x}\ \boxed{3}\ \boxed{=}\ \text{produces the result of } 8^3$$

It should be possible for the second number entered to be a negative number but the first must be positive. The key-stroke sequence

$$\boxed{8}\ \boxed{y^x}\ \boxed{3}\ \boxed{1/x}\ \boxed{=}\ \text{should give the result of } 8^{\frac{1}{3}}$$

$\boxed{\sqrt[x]{y}}$ Calculates the xth *root* of the first number entered (y) provided y is not negative.

These are the most useful, and hence most commonly found, functions on scientific calculators. In addition there will often be a switch allowing the trigonometric functions to be calculated for angles in degrees or radians and possibly even in grads.

It is quite easy for the manufacturer to provide extra 'convenience' functions without too much extra cost, provided the user does not mind having a slightly bulkier machine to accommodate the additional keys. This has led to some rather esoteric functions being available at the touch of a key, but on the whole most are useful, especially to the mathematician. We thus find the following:

$\boxed{\sigma}$ or $\boxed{s}$ *Standard deviation.* It is best to check whether the divisor used is n or $(n-1)$. The Texas Instruments SR 51 in fact allows you to find both since the standard deviation is calculated using $(n-1)$ and the variance is also calculated but uses n. The relationship is $\sigma = \sqrt{\text{variance}}$.

$\boxed{\bar{x}}$ *Arithmetic mean.* Calculates the average of the set of numbers entered.

$\boxed{\Sigma}$ or $\boxed{x_n}$ Used to separate the numbers when entering them for statistical calculations. $\boxed{\Sigma}$ also acts as $\boxed{M+}$ in fact.

$\boxed{x!}$ or $\boxed{n!}$ The *factorial* of the displayed number is calculated: 69 is the largest factorial which can be evaluated and displayed on a machine with an exponent maximum of 10^{99} because $69! \approx 1{\cdot}7 \times 10^{98}$ and $70! \approx 1{\cdot}2 \times 10^{100}$. (The Corvus 500 correctly computes up to 120! but the exponent is reduced by 100 for 70! and over.)

$\boxed{P^n_m}$ Calculates the number of *permutations* of n items taken m at a time: n will be the first number entered, m the second.

C_m^n	Calculates the number of *combinations* of *n* items taken *m* at a time.
sinh cosh tanh	The *hyperbolic functions*. (The *inverse hyperbolic functions* are usually obtained in a similar fashion to the *inverse trigonometrical functions*.)
R→ P→ or R⇌P	These keys are used when converting *rectangular to polar* coordinates and vice versa.
SLOPE INTCP	Used for finding a least squares *linear regression* line for two related sets of data. The *x*, *y* data pairs are keyed in and then by pressing these two keys one at a time the values of the slope and *y* intercept for a linear equation are displayed. It is worth noting that operations such as these may utilize all registers, including user-addressable memories, so a careful check with the instruction manual is required.
∫	*Numerical integration*, which is used to find the area under a given curve by an approximation method such as the trapezium rule.
GAUSS POISS BINOM	*Probabilities* for the Gaussian (Normal), Poisson and Binomial distributions.
χ^2	The *chi-squared* statistic for a set of data may be evaluated using this key.

This is a fairly comprehensive list of the functions which are at present available. Needless to say not all manufacturers offer everything listed and some will even have more, or different, functions to those described. No attempt has been made to include the specialist functions which are found on machines designed for navigators or economists, for example.

It is, however, worth mentioning that several calculators allow the user to select the number of decimal places which are displayed. The key to do this is usually labelled Fix or DSP and should ensure that the last figure displayed is rounded off according to the 'invisible' digit which follows it. Round off means that if the next significant digit is a five or more the value of the least significant digit in the display will be increased by 1; should the value be four or less then the least significant digit displayed is unchanged. On some machines a switch labelled F cut 5/4 may be seen. This allows the user to have a choice of display. At the F position the display will be fully floating with the most significant digits being displayed. The other two positions are used in conjunction with a selectable number of decimal places to be displayed—normally 2 or 4—and will either truncate excess digits or round off the last figure shown.

Another important feature not mentioned so far is the *constant*, which today tends to be 'automatic' on the cheaper calculators and 'non-automatic' on the dearer. On the latter type the constant is distinguished by a $\boxed{K}$ or $\boxed{CONST}$ key which has to be depressed before the constant facility comes into operation. The *constant* means that a number is retained within the calculator for subsequent operations as required. Suppose for example that some conversions of centimetres to inches have to be carried out, then the key sequence on a calculator with an automatic constant would be:

$$\boxed{2}\;\boxed{\cdot}\;\boxed{5}\;\boxed{4}\;\boxed{\times}\;\boxed{3} \qquad \boxed{=} \quad \text{display } 7\cdot62$$

$$\boxed{5} \qquad \boxed{=} \quad \text{display } 12\cdot7$$

$$\boxed{9}\;\boxed{\cdot}\;\boxed{2}\;\boxed{=} \quad \text{display } 23\cdot368$$

and so on. The $2\cdot54\times$ has been retained within the calculator without any effort on the part of the user. If the machine does not have an automatic constant then the key sequence will most likely be:

$$\boxed{2}\;\boxed{\cdot}\;\boxed{5}\;\boxed{4}\;\boxed{\times}\;\boxed{CONST}\;\boxed{3}\;\boxed{=} \quad \text{display } 7\cdot62$$

$$\boxed{5}\;\boxed{=} \quad \text{display } 12\cdot7$$
$$\text{etc.}$$

To be a true constant it must also operate on division without any special procedures being needed. Assume that we wished to convert the above measurements back to inches, then the key strokes on an automatic machine would be:

$$\boxed{2}\;\boxed{3}\;\boxed{\cdot}\;\boxed{3}\;\boxed{6}\;\boxed{8}\;\boxed{\div}\;\boxed{2}\;\boxed{\cdot}\;\boxed{5}\;\boxed{4}\;\boxed{=} \quad \text{display } 9\cdot2$$

$$\boxed{1}\;\boxed{2}\;\boxed{\cdot}\;\boxed{7} \qquad \boxed{=} \quad \text{display } 5\cdot$$

$$\boxed{7}\;\boxed{\cdot}\;\boxed{6}\;\boxed{2}\;\boxed{=} \quad \text{display } 3\cdot$$

It will be seen that a true automatic constant machine retains the first number entered for the multiplication and addition processes but retains the second number for division and subtraction. Some machines only retain the first or second number entered as the constant and this obviously is not so useful.

Many calculators include pre-programmed or 'hard-wired' conversion constants to facilitate the conversion of metric to imperial measure and vice versa. One should check whether the constants used relate to American or U.K. quantities: this will show up with conversions from gallons to litres, for example, where the multiplication conversion factor for U.S. gallons to litres is 3·78541 but for U.K. gallons to litres is 4·54609.

Many problems in trigonometry are set in radians rather than degrees and most calculators offer the choice of entry in one or the other, as well as permitting conversions between the two systems. On the continent the angular measure is the grad—there being 400 grads to a full circle—and conversion to and from this system is sometimes provided.

Another conversion which is becoming more common is that between degrees, minutes and seconds (d.m.s.) and decimal degrees. This will please navigators and school pupils who usually work in d.m.s. since calculators normally accept entries in decimal degrees only. The same format also applies to hours, minutes and seconds and so these can be converted into decimal hours—useful when working out parking meter times or rates of pay. It is very necessary to be able to convert back to d.m.s. after completing a calculation and one should check that the calculator can in fact do this.

The ultimate is surely represented by the Casio AL 10 which claims to be the world's first 'fractional calculator'.

This machine, and some subsequent models, will actually accept entries in fractional notation and display the result as a fraction. (So much for decimalization!)

An example to illustrate this might be:

$$\frac{6}{24} \times \frac{5}{7}$$

The key strokes being

$\boxed{6}\,\boxed{p}\,\boxed{2}\,\boxed{4}\,\boxed{\times}\,\boxed{5}\,\boxed{p}\,\boxed{7}\,\boxed{=}$ display 5⌋28

the display is read as

$$\frac{5}{28}$$

The $\boxed{p}$ key is the 'punctuation' between the fractional parts of the entry. Another depression of $\boxed{p}$ will convert the result into decimal form.

Another unique feature of this same calculator is that it can give the remainders to division sums rather than giving the result as a decimal.

THE CAPACITY OF A CALCULATOR

The *capacity* of the calculator refers to the size of the numbers which it can deal with. If the decimal point moves off to the right too far then *overflow* is said to have occurred—the number has become too big for the machine to display. If the number becomes too small, with the decimal point moving off to the left leaving a string of zeros, then *underflow* has occurred. In both of these cases as soon as the decimal point goes off the display, the machine should lock electronically and display a suitable error signal to indicate the condition. However with underflow some machines simply revert to a zero. (Try squaring the smallest number which your machine can display.) It is this limitation on the size of numbers which makes the simpler machines unsuitable for much scientific or mathematical work at elementary as well as advanced levels. On an 8-digit calculator the largest number which can be handled is 99999999 and the smallest 0·0000001, which is not much use for scientific work.

To overcome this, exponential notation—also known as standard form or scientific notation—has been introduced, which has a range of 10^{-99} to 10^{+99}. An 8-digit machine will thus display numbers either to eight places or in the form of a 5-digit mantissa with a 2-digit exponent depending on the size of the number, e.g. in $1·2345 \times 10^5$ 1·2345 is the mantissa and 5 the exponent. With the so called '5 plus 2' display the 'missing' place is used to space the mantissa and exponent as well as displaying any sign relating to the exponent. Some machines continue to calculate to eight places but will display only the most significant five when working in this mode: the other three may possibly be recalled by depressing the appropriate key. Other common displays are 8 plus 2 and 10 plus 2.

When not in scientific notation, all machines of this type will operate in floating decimal point mode with the decimal point moving so that the most significant digits are always displayed. On some machines it is possible to fix the number of decimal places actually shown by the display. When this happens, the last digit displayed should be rounded off.

Many calculators in fact round off automatically when displaying results. This can be checked for by doing the calculation

$$56 \div 6 \times 6$$

If the result is 56, then round off is taking place. If the result is 55·999999 then it is not: the extra digits have been chopped off.

This rounding off is possible because the calculator is in fact calculating to more places than it is displaying. If we take the SR 51 as a typical example, we find that it does its calculations to thirteen places but only displays ten. To check this do the key strokes

$\boxed{8}$ $\boxed{1}$ $\boxed{1/x}$ the display will be ·012345679

Now subtract ·012345 and multiply by 10000, the new display will be

·006790123

and another four digits have appeared after the 9 and if you count up you will see that the calculation was carried out to thirteen significant figures. The least significant digit in the display is rounded off according to the figure immediately following.

The hidden extra digits are retained in the calculator for use in subsequent calculations. However on the SR 51 if the $\boxed{\text{EE}}$ key is pressed then these internally held extra digits are truncated to make the actual number held correspond to that shown in the display and it is then that number which will be used in subsequent calculations. This can be demonstrated by inserting an $\boxed{\text{EE}}$ key stroke immediately after the $\boxed{1/x}$ key stroke in the previous example, e.g. $\boxed{8}$ $\boxed{1}$ $\boxed{1/x}$ $\boxed{\text{EE}}$, display shows ·012345679. Now subtract ·012345 and multiply by 10000, display is ·006789 and no extra digits are shown.

Just because the calculator displays its result to, say, ten significant figures some people may assume that the final answer which they record will be accurate to that number of figures and so copy down the displayed number without any further thought. They are forgetting that most physical quantities are not even *known* to that degree of accuracy and that the accuracy of the result is limited by the accuracy of the data used to generate it and by the actual calculations involved. We will however assume that you realize this and will not fall into the trap of accepting too many significant figures. Many people, though, find it useful to be able to limit the number of decimal places displayed.

The computational accuracy of any calculator depends on the algorithms which are used to evaluate the functions. One comes across some calculators whose algorithms are suspect, especially with transcendental functions. It is possible for the calculator to build up an error during the course of a complicated calculation. Hewlett Packard give an example of a sine algorithm which involves three divisions, two multiplications, an addition, a subtraction and a square root. These are all treated by the calculator as separate operations and are rounded off in the eleventh place. These round-off errors will accumulate and must be added to any algorithm error. Algorithm errors may arise from the algorithms used to calculate the various functions not being sufficiently exact, perhaps because they have been optimized to reduce either the time taken or the amount of working register space required. The better calculator manuals give details of the algorithm errors which occur on the particular machine. On the whole though these should not worry anyone who is dealing with real data. Round-off errors may show up in relatively simple calculations, especially when computing the difference between two almost equal numbers. A somewhat hypothetical example is shown in the next calculation.

If we wish to evaluate

$$\sqrt{\left(\frac{12398}{435} - \frac{10117}{355}\right)}$$

then the result we get will depend very much on the accuracy to which we work.

$$12398 \div 435 \text{ is } 28 \cdot 50114943 \text{ to } 10 \text{ s.f.}$$
$$10117 \div 355 \text{ is } 28 \cdot 49859155 \text{ to } 10 \text{ s.f.}$$

Rounding off these values and now

working to 4 s.f. the final answer is 0·000000
working to 5 s.f. the final answer is 0·044721
working to 6 s.f. the final answer is 0·050000
working to 8 s.f. the final answer is 0·050567
working to 10 s.f. the final answer is 0·050575

The first two results are not even accurate to 1 s.f. although we are using accurate data and working to 4 or 5 s.f. This kind of problem is of course rather academic with the accurate calculators now used since we generally will work to (say) 8 s.f. throughout.

Another example of interest is the problem of evaluating $\sqrt{(x+1)} - \sqrt{x}$, e.g. $\sqrt{97865432} - \sqrt{97865431}$, as accurately as possible.

Using an 8-digit calculator we get:

$$9892 \cdot 6959 - 9892 \cdot 6958 = 1 \times 10^{-4}$$

and there is only 1 s.f. in the answer!

However $\sqrt{(x+1)} - \sqrt{x}$ is the same as $1/(\sqrt{(x+1)} + \sqrt{x})$ so if instead we calculate $1/(\sqrt{97865432} + \sqrt{97865431})$ we should get the same result.

In this case the answer is $5 \cdot 0542340 \times 10^{-5}$—a very different result and containing 8 s.f. all of which are correct! Thus the *same* data on the *same* calculator can yield *very different* answers depending on the procedure used.

The basic problem here is the cancellation of many leading (significant) figures when two very similar numbers are subtracted.

With $\sqrt{(x+1)} - \sqrt{x}$ we are subtracting two similar numbers (if x is large) but with $1/(\sqrt{(x+1)} + \sqrt{x})$ there is no subtraction involved at all, and so no cancellation and loss of significant figures occurs.

A further example is included in Chapter 6 (page 116).

Good quality machines are unlikely to give the user any problems regarding machine accuracy. For their SR 52 Texas Instruments state that the 12-digit representation of π held as a constant exceeds the exact value of π by less than $2 \cdot 07 \times 10^{-13}$, i.e. less than $6 \cdot 6 \times 10^{-12}$ percent. This represents an error equivalent to about 0·0025 mm in computing the equatorial circumference of a sphere the size of the earth, given its radius, a precision adequate for most purposes.

3. Inside the calculator

Let us start with the integrated circuit, or 'chip', which contains the brain of any calculator. This is a plastic or ceramic package measuring some $3 \cdot 7$ cm $\times 1 \cdot 4$ cm $\times 0 \cdot 5$ cm with about twenty legs or pins sticking from it. Inside this package will be the LSI (large-scale integrated) chip itself, which is about $0 \cdot 5$ cm square with connections being made from it to the pins by means of very fine wires.

Each chip carries on itself the equivalent of thousands of components such as transistors, resistors, and diodes. These are all connected to form the basic logic gates which perform all the necessary logic functions. This book is not the place to go into the theory of circuit logic but a simple example may suffice to make things clearer.

The lamp in the circuit (fig. 3.1) will light if either switch A or switch B is depressed, whilst in the second circuit (fig. 3.2) the lamp will only light if switch A and switch B are both depressed.

The circuit in fig. 3.1 acts as an OR gate, the one in fig. 3.2 as an AND gate. It is the electronic equivalents of these gates and others like them which are used as the basic building blocks in calculators. The gates most commonly used are NAND and NOR gates, the N at the beginning standing for Negation (so NAND is 'not AND'). In our example a NAND gate would be one where the light went off when both switches were closed and was otherwise on. The electronic gates can be made from a combination of resistors, transistors and diodes and today are

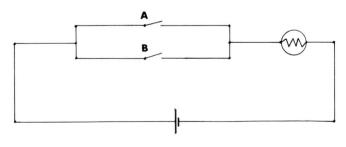

Fig. 3.1   OR gate.

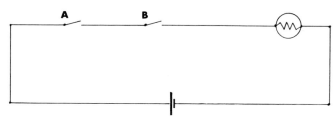

Fig. 3.2   AND gate.

manufactured using the metal oxide semiconductor technology (or MOS). Most people will be familiar with the transistor, which is commonly a bipolar device, so called since the current carriers are both negative (electrons) and positive (holes) travelling in opposite directions. Integrated circuits (ICs) can be fabricated using bipolar transistors but have several disadvantages, not least of which are the limited number of transistors which can be packed onto a given sized chip and their high power consumption. A better alternative to the bipolar device is the field effect transistor—or FET—which, by suitable modification, can form resistors or diodes. Bipolar technology may regain some favour with the introduction of the I²L technique, also known as 'injection logic'.

The IC designer, using a computer, decides which logic elements are needed for the functions required on the calculator together with their necessary connections. From this a circuit diagram can be obtained which a computer can rearrange to optimize the layout on the chip. The transistors of the IC are built up in layers using a photographic process and each layer will require a separate negative, or mask, as it is called. The masks, five or so in number, are drawn by the computer from the final circuit diagram and define not only the areas which are to be transistors but the various interconnections as well.

The original masks drawn by the computer plotter are about a thousand times the size of the final chip. Once they have been checked for correctness they are photographically reduced to the size of the chip itself.

Each chip is just part of a single slice of an extremely pure silicon crystal which has been produced in a cylindrical bar form by a pulling process from a bath of molten silicon. Each silicon slice—which may be between 5 cm and 12 cm in diameter with a thickness of some 0·2 mm—can produce several hundred separate chips. Thus the masks produced are repeated side by side using a step-and-repeat camera until the final master mask covers the whole area of the silicon slice.

The slice, which by now has had a silicon oxide coating applied, is coated on its top surface with a 'photoresist' (a chemical which hardens under ultraviolet light). The slice is then exposed to ultraviolet light shone through the first mask and some areas of the photoresist harden, corresponding to areas on the mask. The parts where the resist has not hardened are removed, leaving 'windows' through which the oxide layer is thickened within carefully controlled limits. Then another coating of the photoresist is applied and the second mask is exposed. Through the resulting windows, polycrystalline silicon is deposited which is then oxidized. More photoresist is applied and the third mask exposed. This time a boron impurity is diffused into the windows which result, to produce the electron deficient centres necessary to give semiconductor properties. Other steps in the process are concerned with an aluminium metallization pattern which forms the connections, and the formation of a passive dielectric layer. Finally a protective layer of glass is deposited over the whole slice.

Remember that all this is being done to a hundred or so silicon slices each with about a thousand chips on their surface. All the work is carried out under very clean conditions and an integrated circuit may take a month to go through all the various processes, most of which are automated. Finally, special machines lower probes onto each individual chip to check its electrical properties.

In some respects we have started at the wrong end of the story, since before the chip can be designed, the tasks it has to carry out need to be defined in terms of the logic required for the functions which the manufacturers have decided should be available on the keyboard. Before the logic can be worked out, the algorithms which will compute the functions have to be devised. Algorithms are the routines which the calculator follows in order to produce the value of a particular function. The logic circuits themselves are restricted to addition-type routines owing to the logical combinations which they can carry out using variations of the NAND and NOR gates which we have met. This restriction means that every function has to be presented in a series of addition-type processes. Another slight complication is that the decimal numbers with which we are familiar have to be converted into another form which is more acceptable to the logic circuits. This is a variation of the binary system, which most readers will have come across. In binary notation each digit represents a power of two; thus only two figures are required, a 0 and a 1.

For example the binary number 101001, when read from right to left indicates that there is a '1' in the first place, there is no '2', no '4', an '8', no '16', and a final '32'. The decimal equivalent is $1+8+32$ which equals 41. One adaptation of pure binary is *binary coded decimal* (BCD), in which each separate decimal digit is represented by a four-bit code. (A bit being one binary digit.) Taking our previous example, 41 consists of two decimal digits, a four and a one. Encoding the 4 we get the binary 0100 (no 8, a 4, no 2 and no 1) and similarly encoding the 1 we get 0001 (no 8, no 4, no 2 and a 1). Arranging these BCD numbers according to their decimal place we get the final form 0100 0001. You will note that to encode a decimal number, one does not use all the possible combinations available in the four-bit code. A 1 in each place —1111—would represent 15 in the equivalent decimal form, whilst the highest number we require is nine (a ten must appear as a 1 in the next place to the left). These extra combinations—the numbers 10 to 15— can be used as instructions to tell the calculator what to do, provided they can be separated from the data.

The advantage of using binary coded information instead of decimal is that only *two states* are required to represent each digit. This is easily achieved in electronic circuitry by using On, or a higher voltage, to represent 1; and Off, or a lower voltage, to represent 0. There is no need for any intermediate values.

One very important circuit block used in calculators is the shift register, normally called the memory. This too is constructed from a combination of gates which are connected to form bistables or flip-flops, as they are sometimes called. Each bistable can have only one of two possible states at any particular instant. Thus in one state it can be used to indicate a 1 and in the other a 0. These states are quite stable and the bistable will remain in whichever state it has been set until changed by an external signal. One way of visualizing a bistable is to imagine it as a toggle which can be either up or down, up being equivalent to a 1. Thus our number 41 would be represented as in fig. 3.3.

Fig. 3.3 A serial 'toggle store' containing the number 41. An UP toggle represents a high whilst a DOWN toggle represents a low.

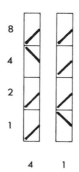

Fig. 3.4 A parallel 'toggle store' containing the number 41.

This shows how the bistables would be arranged if the data were to be stored in series. However the working registers and memories of a calculator are often organized to deal with the data presented in parallel. If this is the case the diagram changes to that in fig. 3.4.

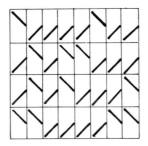

Fig. 3.5   A toggle register or memory containing an 8-digit number—the number is 97064813.

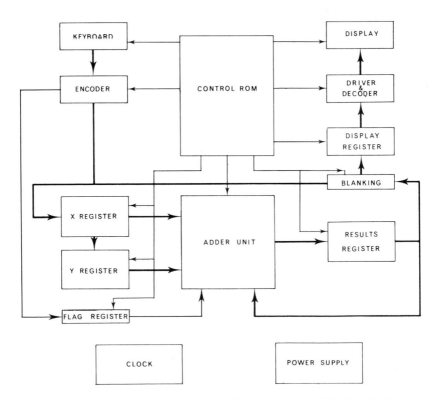

Fig. 3.6   This diagram attempts to show how the various 'blocks' which go to make up a complete calculator are organized. The clock and power supply are common to all.

Each column is then used to store one decimal digit and so for an 8-digit number at least 8 separate columns involving a total of 32 separate bistables are required (fig. 3.5).

The difference between a working register and a memory lies in the ways in which they are controlled. A working register is usually controlled by the calculator itself and responds to the instructions of the algorithm: a memory is at the disposal of the user to store numbers when required.

The working of a calculator can best be explained with reference to a simplified block diagram of how its various parts are connected (fig. 3.6).

We shall use this to illustrate how a simple algebraic machine processes various functions. The actual wiring required for a real calculator is shown in fig. 3.7. The keyboard is the input interface between the user and the calculator and consists of a number of switches located at the intersections of a matrix. A 4×4 matrix can therefore service 16 keys, a 4×9 matrix can service 36 keys and so on. One set of matrix lines is designated scan lines whilst the others are input lines, so each switch provides the only possible connection between a scan line and an input line. The scan lines are activated in turn and, should a key be depressed at the moment of activation, then a pulse will be transmitted into the appropriate input line. Decoding circuits can thus identify which key has been pressed and pass the information on. Allowance has to be made for electrical noise, which could produce a meaningful but spurious input signal. The calculator verifies the input by checking if it is still present on the next scan (≈ 2 milliseconds later). If it is, it will be accepted. Key bounce is another problem and is caused by the fact that when a switch is closed or opened, the contacts oscillate and this can cause unwanted extra inputs if an 'antibounce' program is not included.

After a digit has been keyed in, it is decoded and translated into a BCD form which is then stored in the X register and also in the display register, from whence it is decoded and shown in the actual display in ordinary decimal form. (Our hypothetical calculator has separate X and display registers, although this may not be the case on the simpler machines.) The next digit to be entered goes through the same processes ending up by being shown in the display and also 'pushing' the first digit along one place in the shift register.

Any further digits which are entered to form this first number will be displayed and stored in the X register sequentially. The maximum number of digits which can be entered to produce a number is governed by the capacity of the registers. Usually any extra digits are not recorded.

The next step in the calculation is to enter the operator, e.g. $+$, which is decoded and recognized as an operator and stored in a separate register—the flag register—in a BCD form. The next number is then

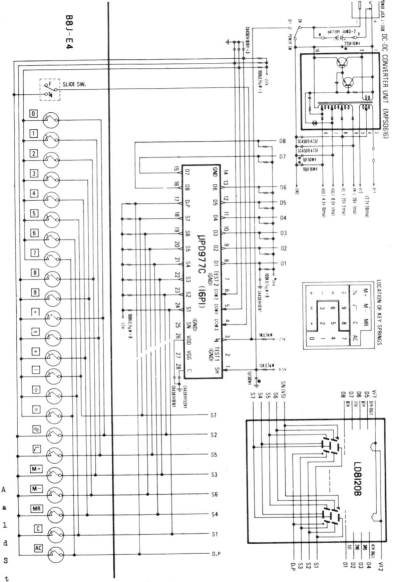

Fig. 3.7 A circuit diagram of an 8-digit, digitron, calculator with memory (the Casio 8A). The keyboard and display will be seen to be connected to the same scan lines, S1–S7 and DP. These lines are each energized in turn and if a key contact has been made then it will be detected by one of the keyboard input lines which are connected to pins 4, 5 and 6. Similarly the appropriate segment of each digit, D1–D8 will be illuminated according to which digit has been selected by the control circuitry within the chip as each scan line is energized. Notice also that a DC–DC converter is required to provide the necessary voltages for the I.C. and display and that the internal battery is disconnected when the mains unit is plugged into the power jack.

entered and immediately the first digit is keyed in, the contents of the X register are transferred to the Y register leaving the X register empty and so ready to accommodate the new number.

Suppose we are doing a relatively simple sum, say 3+6, on an algebraic machine. A ⟦3⟧ is keyed in followed by the ⟦+⟧ and then the ⟦6⟧. Finally the ⟦=⟧ key is depressed to complete the calculation. It is this equals key depression that starts the calculator actually calculating, since up to then everything—numbers and operator—have just been stored in various registers. But now the control ROM (read-only memory) can begin the process of combining the two numbers according to the operator key which has been depressed. Within the control ROM all the instructions which the calculator will need to carry out have been pre-programmed in a BCD form into registers whose contents cannot be destroyed when power is switched off. They can be read but cannot be altered in any way, hence the name Read-Only Memory.

The two numbers entered in our sample problem were 3 (0011) and 6 (0110). These are fed in parallel into the adder unit which will put the answer of 1001 (9) into the results register. Note that during the addition process 'carrys' may be generated and that these have to be added into the next place. For example, in binary $1 + 1 = 0$ carry 1. These carrys may be remembered in small registers within the adder unit itself. During the actual calculation the results register should be disconnected from the display register via the blanking circuitry so that the display does not show the calculation taking place. This does not always happen on some of the cheaper models. Once the calculation is finished the control circuitry allows the results register to be copied into the display register so that the answer can be seen.

The process of adding all four digits at once is known as 4-bit parallel addition. The units are dealt with first followed by the tens, then the hundreds, and so on. You may, though, be wondering what happens if the sum of the units, say, exceeds 9, as in the example: 85+29.

Converting this to BCD we have 1000 0101+0010 1001. Dealing with the *units* first and adding, following conventional rules, we get:

$$
\begin{array}{r}
0101 \\
1001\,+ \\
\hline
1110
\end{array}
$$

which, if decoded, is 14 and hence too big for the units column. So somehow a carry needs to be generated and this is done in the calculator by adding a 6 to the resulting number whatever it is:

$$
\begin{array}{r}
1110 \\
0110\,+ \\
\hline
1\quad 0100
\end{array}
$$

The carry 1 is remembered in an internal register whilst the 0100 is passed on to the results register. Now the tens can be dealt with and we have:

$$\begin{array}{r} 1000 \\ 0010 \\ \underline{1+} \\ 1011 \end{array}$$

decoded we find this is eleven so once again the control circuitry will sense that it is larger than 9 and add a 6:

$$\begin{array}{r} 1011 \\ \underline{0110+} \\ 1\quad 0001 \end{array}$$

and the answer is passed on to the results register, which now holds 0001 0100. Since there are no further numbers in the X or Y registers, the carry 1 is shifted into the results register giving a final answer of 0001 0001 0100 or 114. The adder unit itself is pre-programmed to add the 6 and this is known as a hard-wired function.

Subtraction is relatively easily carried out using the adder, provided the digits which are to be subtracted are first converted to their *complement* form. For example, 1001 would become 0110 when complemented. Suppose we wished to work out the answer to $9 - 5$, then converting into BCD we get $1001 - 0101$. The 0101 is complemented to become 1010 which is *added* to the 1001 like this

$$\begin{array}{r} 1001 \\ \underline{1010+} \\ 1\quad 0011 \end{array}$$

Note that a carry 1 has been generated and this is added back onto the answer at the least significant end

$$\begin{array}{r} 0011 \\ \underline{0001+} \\ 0100 \end{array}$$

which, when decoded, gives the correct answer of 4. This may seem a long-winded way of doing subtraction but electronically it is quite easy.

So far we have only dealt with integers. What happens to any decimal part? Well, there are several ways of locating the decimal point within the various registers. One method uses a non-operational BCD code, i.e. a number between 10 and 15 which is generated when the decimal point key is pressed, and stored with the digits in the registers. Another method counts the number of digits entered before or after the decimal point and this number is stored in a separate register. Whichever method is employed the principle used when working with decimal numbers is to shift the contents of one register with respect to the other until both decimal points are 'lined up'. During this lining up process the least significant digits of the smaller number are dropped off as they reach the right-hand end of their register. Once the two decimal points coincide, the digits from both registers are fed into the adder. If the final result is greater than the capacity of the results register, for example if a carry 1 exists in the adder which cannot be accommodated in the results register, then the condition is sensed and an *overflow* indication should be given. Similarly an *underflow* indication might be given if, during the lining up process, one of the registers drops off all its digits, meaning that it is impossible to line up the two decimal points. $(9999999 \cdot + 0 \cdot 0000001$ for example.)

The other instruction which needs to be stored is the change-sign command and this may be stored as either a non-operational BCD code or in a separate small register associated with the relevant working register.

By suitably extending the decimal-point location-counter idea, the calculator can be made to cope with exponent notation, giving an exponent range, provided the sign is suitably encoded, from 10^{-99} to 10^{99}. Most calculators without exponent notation will work in floating decimal point mode, in which only the most significant digits are displayed. You will see that this is a consequence of using the decimal lining up technique. Many modern calculators work in either floating-decimal-point or exponent notation depending on the size of the numbers entered, the answer being displayed in the more appropriate form. Others immediately 'normalize' (convert to exponent notation) the number just entered when any function key is pressed, so that the mantissa is between $1 \cdot 0000000$ and $9 \cdot 9999999$.

To illustrate this let us suppose that $34567 \cdot 89 \times 10^8$ had been entered; this would be normalized to become $3 \cdot 456789 \times 10^{12}$.

Before addition or subtraction can be carried out in exponent notation, the contents of the two exponent registers have to be equal and so decimal points may have to be shifted within the working registers to achieve this before the mantissa can be added. Thus the calculator, if presented with the problem of adding $1 \cdot 2345 \times 10^6$ to $6 \cdot 789 \times 10^8$, will adjust the smaller number to become $0 \cdot 012345 \times 10^8$ and then add to give an answer of $6 \cdot 8013 \times 10^8$. Note the loss of digits from the smaller number in this process.

If we are to keep this explanation simple, we had better leave addition and subtraction now and turn to the other two arithmetic operations, multiplication and division. As has already been said, the calculator can only add, so these processes have to be turned into addition. The routines which will do this—the algorithms—are stored in the control ROM. The simplest way of getting an answer to 53×84 would be to add 53 to itself 84 times (or vice versa). Quite simple to achieve but very slow. A faster method is to use a 'shift and add' technique very similar to that used on the old mechanical calculators. To multiply 53 by 84 you would firstly add together four 53s to give a partial product of 212, which is then shifted one place (in decimal notation) to the right. Now eight 53s are added together to give another partial product of 424 which is not shifted. Rather than illustrate this in BCD let us stick to decimal, which gives us

$$
\begin{array}{r}
212 \\
424\ + \\
\hline
4452
\end{array}
$$

It is worth noting that at least thirteen separate passes through the adder are involved for this simple multiplication—four times for the units, eight times for tens and once for the partial products. The reason why the first partial product is shifted to the right rather than the second being shifted to the left is simply that the former is easier electronically. A similar approach can be used for the exponent parts of numbers.

Division on the other hand is done as a repeated subtraction routine after the necessary complements have been produced. The principle is easiest to explain using an example, which once again will be in decimal rather than BCD so as not to complicate the issue too much. Suppose we have to divide 144 by 32. The BCD equivalent of 144 will be in the Y register whilst the BCD equivalent of 32 will be held in the X register. The 32 is subtracted from 144 as many times as is possible until the contents of the Y register are as small as possible without going negative. (In fact the contents do go negative and this is detected and the number being subtracted is added back once to make it positive; but let us forget this complication for now.) The number of subtractions is recorded in a register. The remaining contents of the Y register are multiplied by 10 and the subtraction is repeated until the remainder is just non-negative, the number of subtractions is recorded and the remainder is multiplied by 10 . . . and so on until a result is obtained or the capacity of the results register is exceeded. In the example we chose the various stages are:

$$144 - 32 - 32 - 32 - 32 \text{ (four times) remainder } 16$$

$$16 \times 10 = 160$$

$$160 - 32 - 32 - 32 - 32 - 32 \text{ (five times) remainder } 0$$

Hence the answer is 4·5.

By now you will have started to realize that the calculator is not quite so simple as it at first appears! A lot of time, money and thought has gone into optimizing the algorithms used for these relatively simple arithmetical processes. The intermediate numbers may pass through the calculator many times with partial answers being stored and recalled as required by the program contained in the control ROM.

The matter becomes even more difficult once the scientific functions are included in the keyboard instruction set and it is very difficult to find which algorithms are actually used. One major manufacturer refuses to divulge any information about the algorithms used in his machines since he feels it could give his competitors useful information. One realizes why this is such a sensitive area if one compares the times taken by machines from different manufacturers to work out the same function. Obviously they do not all use the same algorithms and some are much slower than others, although just as accurate. The only manufacturer who seems to have published anything in this area is Hewlett Packard who use a 'pseudo division' and 'pseudo multiplication' method. This is not the place to go into great detail of this complex technique so the relevant references are supplied [J. E. Volder (1959) and J. E. Meggitt (1962)]. Very simply, though, the basic division routine already mentioned is modified by the addition of an extra register which is used to change the value of the divisor after the divisor has shifted to the left and before the next subtraction takes place. Initially the divisor might be A, and A would be subtracted as many times as possible from B leaving a remainder of C which is shifted one place. Now $A + A \times 10^{-1}$ is subtracted. For the next shift the divisor becomes $A + A \times 10^{-1} + A \times 10^{-2}$ and so on. After this it gets rather difficult to explain simply but the net result is that these algorithms can produce logarithms, powers, tangents, inverse tangents and exponentials.

It is probable that this algorithm is not peculiar to Hewlett Packard and other manufacturers most likely use similar routines. One of the most important derived functions is the natural logarithm, because once this is known it can be used to calculate powers and base ten logarithms using normal arithmetical rules. On the CBM SR 1800 the display shows the natural logarithm of the number entered (y) when the $\boxed{y^x}$ key is pressed, and before the power (x) has been entered, indicating that y^x is really evaluated as $e^{x \ln (y)}$. Once the control ROM has been suitably programmed, quite complex algorithms can be performed using iterative techniques.

For another example Casio have kindly supplied the following information on the algorithm used to calculate square roots on the Casio Memory 8 Calculator.

The square root is extracted by what is basically a subtraction process in which the subtrahend, i.e. the number being taken away, is increased by 10 after each subtraction. This new subtrahend is then subtracted from the result of the previous calculation unless it is too big, in which case a reformulation is necessary before continuing. The easiest way to follow the operation of this algorithm is by an example. Let us suppose we wish to find $\sqrt{5 \cdot 3}$.

Firstly the number (5·3)
 is multiplied by 5 $5 \cdot 3 \times 5 = 26 \cdot 5$
Now 5 is subtracted   5 −
           ——
Remainder...................... 21·5
10 is added to the last
 subtrahend (5) and the
 result is subtracted
 from the remainder .... 15 −
           ——
Remainder...................... 6·5
10 is again added to the 25 (too big) *1st estimate 2·5*
 last subtrahend (15)
 but the result (25) is
 too big, so a reformu-
 lation is necessary. The
 remainder (6·5) is mul-
 tiplied by 100 .... { 650

The subtrahend
 is 'opened' by   Reformulate }
 inserting a
 zero to the
 left of the
 units digit .... { 205 −
           ——
Now the process recom-
 mences 445
          215 −
          ——
          230
          225 −
          ——
           5
          235 (too big) *2nd estimate 2·35*

Reformulate { 500
       2305 (still too big) *3rd estimate 2·305*
       ——
Reformulate { 50000
      23005 −
      ——
      26995
      23015 −
      ——
       3980

$$23025 \quad \text{(too big)} \qquad \textit{4th estimate } 2\cdot3025$$

$$\text{Reformulate} \begin{cases} 398000 \\ 230205 - \end{cases}$$

$$167795$$

$$230215 \quad \text{(too big)} \qquad \textit{5th estimate } 2\cdot30215$$

and so on

This would once again be processed by multiplying by 100 etc. and the subtraction continued. However for the purposes of this example this is a convenient place to stop, noting that an estimate of the square root of 5·3 has been obtained to 6 s.f.: 2·30215, which is the last value of the subtrahend. The correct value of $\sqrt{5\cdot3}$ is 2·302173 to 7 s.f. and repeating the process through one further cycle would yield 2·302175. Note that the position of the decimal point is organized by the control ROM, and the process is best considered as applied to numbers in the range $1 \leqslant x < 100$, as with square root tables.

Using this algorithm, the reader may like to work out the square roots of some other numbers, including perfect squares.

Pi, π, is usually a hard-wired function, its value being kept permanently stored in a separate register rather than being calculated each time it is required—which can be often, especially when radians are being converted to degrees or vice versa, a feature found on many machines.

One big advantage for the calculator manufacturer is that once he has his adder unit and its various storage registers, collectively called the *arithmetic logic unit*, it is relatively easy to substitute another, different, control ROM to provide a new set of keyboard functions. This is what Hewlett Packard have done with their HP 21/22/25 range. Here the ROM comes as an 18-pin plastic package organized internally as 1024 'words' each ten bits long. It is interesting to speculate what might come next in this range, since the arithmetic logic unit can directly address up to four separate ROMs and only two are used at present.

The part of the calculator we have not dealt with so far is the *clock* which generates a square-wave signal which serves to keep all the activities of the various subsystems in step.

One of its major jobs is to move the contents of the shift registers along and the time taken to move a digit from the first location in a register along the register and finally out is known as an *instruction cycle*. An addition, for example, may occupy 300 instruction cycles and take about 0·05 sec to perform.

The display, which may either be a red LED or green digitron, is multiplexed. This means that each digit is only energized for a fraction of the time. In the HP 21, any particular segment is energized for only 1 % of the total time. No flicker is seen, owing to the high frequency rate of the clock, which controls this operation too. Special transistors have to be provided on the integrated circuit to handle the relatively large current needed by the display. In many cases these are external to the main chip though CBM have the honour of actually being the first manufacturer to produce a one-chip calculator. Admittedly this is only a 'simple' calculator and the displays used with it have to be specially selected but even so it represents a big technological advance, which others have followed.

The power for all the different parts of the calculator is derived from disposable batteries, rechargeable cells or the mains. Whichever method is used the basic voltage of about 1·5 V per cell or a.c. mains voltage has to be changed to give supplies of $+6$ V and -9 V (these values depend on the make of chip used). This conversion is carried out by changing the d.c. into a.c. by means of a small vibrator, transforming the a.c. to the required voltages, and then rectifying it. With a radio one can often pick up this vibrator frequency, together with the clock frequency, and these electromagnetic transmissions have occasioned some alarm since it was thought that they might interfere with other equipment, especially on board aircraft.

Having looked at how a calculator is organized electronically, let us now delve into the organization of its logic. We shall first investigate a typical machine with algebraic logic and then extend this to cover the Texas Instruments Algebraic Operating System, AOS. For Reverse Polish notation, RPN, we shall choose the HP 21 as a representative machine.

An algebraic calculator has two working registers—called X and Y—and perhaps a user-addressable memory. There may be other registers within the machine which hold intermediate results, and the X register itself may be made up of others which work so closely together that they are indistinguishable to the user. The X register in its entirety is the input register, and the value held in it is always the operand for single-variable functions such as the trigonometric functions, logarithms, reciprocals, squares and square roots. After these functions have been processed, the result is returned to the X register and the contents of the Y register should be left unaffected. We say 'should' because some calculators cannot directly handle problems such as sin 30 + cos 30; it would appear that the first result in this type of calculation is lost when the next function is keyed in because there are insufficient registers to cope with the computation of the transcendental functions if they occur sequentially. When doing multiplication or division or any two-variable operation, the first operand is held in the Y register and the second in the X register. The Y register also acts as the 'store' for the result of previous calculations which form part of a longer calculation. (But see the note, nine lines above, for cases when this breaks down.)

Texas Instruments machines are slightly more sophisticated if the Algebraic Operating System is incorporated. This system deals with functions according to their order in the algebraic hierarchy; thus multiplication and division are performed before addition and subtraction. In order to achieve this, an extra register, the Z register, is provided, which acts as a cumulative register for the results of addition and subtraction processes. The following example should clarify how the three registers on Texas Instruments machines operate to provide their distinctive 'sum of products' logic. If the problem $5+6\times9-8$ is to be carried out, the contents of the various registers will be as shown in fig. 3.8.

Entry	5	+	6	x	9	–	8	=
X	5	5	6	6	9	59	8	51
Y	0	0	0	6	6	6	6	6
Z	0	5	5	5	5	59	59	0
Flag		+	+	+ & x	+ & x	–	–	

Fig. 3.8 Contents of the X, Y and Z registers on a calculator with algebraic hierarchy during the calculation $5+6\times9-8$.
Note that at one point there are two pending operations being held in the flag register.

Depressing the [+] key shifts the 5 into the Z register, allowing the 6 to go into the Y register and the 9 into the X register. (If the [x⇌y] key is pressed after entry of the 6 has been made it will show that there is no number in the Y register.) The depression of the [–] key completes the two pending operations in the order 6×9 to give 54 followed by the addition of the 5. The display then shows the intermediate result of 59. If the 8 is now entered and the [=] key pressed the final answer will be displayed. The [=] key in fact completes any operation outstanding, in the correct hierarchical order, as well as clearing the Z register.

Memories on algebraic logic machines, or for that matter on Reverse Polish machines, are usually completely separate from the working registers in that their contents are not affected by the intermediate results of internal machine computations. The exception may be with the statistical functions. It is worth noting that, before starting any process which does use the user-accessible memories, these memories must be cleared or else their contents may interfere with the data being stored there. Many algebraic machines now offer parentheses, or brackets, which enable pending operations—and their results—to be retained until required by the calculator in its solution process. The inclusion of parentheses obviously demands extra registers to provide the additional storage. In most cases the parentheses registers are completely separate from user-accessible memory registers.

Hewlett Packard machines use the Reverse Polish logic form of entry (RPN) which in some instances can simplify the working by eliminating the need for parentheses. In these machines there are four registers, called X, Y, Z and T, which form the operational stack. The X register acts as the display register showing what has been keyed in. To enter the first number into the stack the ⊡enter⊡ key is depressed so that the number in register X is transferred into Y. As a consequence of this the contents of Y are copied into Z, the contents of Z goes into T and the contents of T are lost. Operations only take place between the contents of the X and Y registers, with the Z and T registers acting as holding stores for earlier results or for entries which may be needed later. It is useful to see the contents of the stack during the working of a problem such as $(a+b)\times(c-d)$. Rewriting this in Reverse Polish we get $ab+cd-\times$.

When this is loaded into the machine an ⊡enter⊡ key stroke will be needed after a and again after c to separate out the adjacent numbers. An example is given in fig. 3.9.

T	0	0	0	0	0	0	0	0	0
Z	0	0	0	0	0	15	15	0	0
Y	0	7	7	0	15	6	6	15	0
X	7	7	8	15	6	6	4	2	30

Entry	7	enter	8	**+**	6	enter	4	**—**	**×**

Fig. 3.9   Contents of the stack on a calculator which uses Reverse Polish logic when doing the calculation $(7+8)\times(6-4)$. Note that operations are executed immediately they are keyed in and how too the numbers move within the stack.

In this system, since the stack is a 'last in, first out' memory the operands required next by the user are always at the bottom of the stack. In addition to the ⊡enter⊡ key a ⊡roll↓⊡ key is provided so that the contents of the stack can be reviewed or the whole stack moved round one place at a time if required. Four consecutive operations of the ⊡roll↓⊡ key return the registers to their original state with no loss of data.

As with the SR 51 the single-variable function keys operate only on the displayed number.

In practice more registers than those described here are required to hold all the partial results which will be generated by the full range of operations described earlier.

In this brief look at how calculators work you should have gained some insight into the complexity of the programming required to implement the various functions. The story does not end here. We have covered very briefly the electronics and logic but what about the design of the case, which has to be slim enough to fit into the hand and yet sturdy enough to withstand the knocks of everyday use? Then there is the power supply, the display, the keyboard and even the marketing. All the time the designers are trying to cut down production time—which costs money—by squeezing more of the components onto the chip, using clip-together cases, and so on.

From the outside the latest generation of calculators may look very similar to the last but inside this is certainly not so. Calculators are today 'state of the art' pieces of equipment incorporating the very latest technological knowledge and this is likely to continue, especially with the present competition, where manufacturers are vying with each other not only in terms of price per function but in the technology which is needed to bring the machines onto the market at all.

4. The basic functions in action

The fractions, or rational numbers, can be considered as ratios of integers, e.g. $\frac{2}{3}$, or as the result of dividing one integer by another integer, e.g. $2 \div 3$. On the calculator only the second of these representations is normally feasible. (Some Casio machines however can work in fractions.) The only way to enter a fraction is to perform the relevant division and thereby produce a *decimal fraction* (usually called simply a decimal). The division itself can be performed on the machine or the decimal fraction can be entered directly. Of course the representation of a fraction as a decimal is often entirely acceptable, e.g. $\frac{2}{5}$ becomes ·4 exactly. However, on an 8-digit machine $\frac{1}{3}$ becomes ·33333333 which is only an *approximation* to $\frac{1}{3}$, and 125/1024 becomes ·12207031, again only an approximation. There is nevertheless a difference in kind between these two cases: $\frac{1}{3}$ can *never* be exactly expressed as a decimal on a calculator, no matter how many digits are taken, because $\frac{1}{3} =$ ·33333333 . . . . . without end; $125/1024 = $ ·1220703125 exactly so a 10-digit machine could fully represent 125/1024.

Given a calculator with an 8-digit display, say, what happens if we compute $1 \div 3$ and then multiply the result by 3 again? This will depend on the particular machine. If the machine only works to the 8 digits which it displays the result will be $(1 \div 3) \times 3 = $ ·99999999, but a more sophisticated machine which carries hidden decimal places and rounds off for the display will produce 1. For example if the machine actually works to 10 figures it will calculate $1 \div 3 = $ ·3333333333 and $(1 \div 3) \times 3 = $ ·9999999999. Now rounding off to eight figures for the display gives 1·0000000 and the trailing zeros will be suppressed to show 1.

There are drawbacks to both systems. On the one hand the simpler machine keeps giving the 'wrong' answer, or less accurate answers. On the other hand the sophisticated machine keeps hidden part of the decimal number and the user may not be aware of what is hidden. The *same display* multiplied by the *same number* can give different results, depending on what is hidden.

Generally speaking the user will not need to worry about any extra figures carried. They will simply give greater accuracy to the working. It is nevertheless useful to be able to access the hidden digits—and so make use of the full accuracy of the machine—and the next section will introduce this idea.

RECURRING DECIMALS

Suppose we want the decimal equivalent of 1/70. On a 6-digit display (8-digit working) calculator $1 \div 70$ is displayed as ·014286 where the last digit is subject to rounding off. How can we get more digits in the decimal expansion of 1/70? Multiplying the displayed number by 10 produces ·142857 and because we have eliminated the leading 0 which takes up one place we have gained a digit. Multiplying again by 10 yields 1·42857 and so no extra information has been gained. However by subtracting 1 we eliminate another leading figure and make room for one more at the other end, giving ·428571. Now multiplying by 10 and subtracting 4 gives ·285714. This is as far as we can usefully go because only (suppressed) trailing zeros will occur if we repeat the procedure further as we have accessed all the 8 digits. We conclude that

$$\frac{1}{70} = ·014285714\dots\dots$$

and a reasonable guess is that we have a repeating pattern, signifying a recurring decimal, so we conclude that

$$\frac{1}{70} = ·0\dot{1}4285\dot{7}$$

with a repeating period of length six digits.

Many recurring decimals have much longer periods and so defeat the above method of investigation. For example

$$\frac{1}{19} = ·\dot{0}5263157894736842\dot{1}$$

and not many machines work to the 20 or more figures which are needed to discern this recurring pattern!

However it is not too difficult to devise a scheme to generate an indefinite number of decimal digits, suitable for any calculator, using the principle of long division, and we shall return to this later in the chapter.

Something to do

Exercise 4.1   Investigate the accuracy to which your machine works by calculating $1 \div 7$ and seeing how many digits of ·1428571428 .... you can uncover.

PERCENTAGES

Whether or not your calculator has a 'percentage' key is of little importance when it is recalled that a percentage is simply a fraction of 100,
e.g. 30 per cent (30%) is 30 hundredths = 30/100 = ·3. Thus to calculate
30 per cent of £256.00 any of the following procedures will do, working
from left to right:

  (i) $30 \div 100 \times 256 = 76·8$
  (ii) $30 \times 256 \div 100 = 76·8$
  (iii) $·3 \times 256 \qquad = 76·8$

Notes

  Method (i)   may *appear* ambiguous—it is $(30 \div 100) \times 256$ rather than
$30 \div (100 \times 256)$—but in practice calculators are always built to work
such expressions out in the desired left-to-right order.

  Method (ii)   avoids any appearance of ambiguity and in simple
problems the last division by 100 can often be omitted and the correct
answer written down straight away as it merely involves shifting the
decimal point two places to the left.

  Method (iii)   is to be preferred in simple cases when the user is sure
that he will not make an error in working out the percentage. However
it is easy to put ·3 for 3% by mistake. Also if a very accurate calculation
of, say, $16\frac{2}{3}\%$ of P is required it would be preferable to work out
$1 \div 6 \times P$ rather than to enter ·16666667 and multiply by P, particularly
if extra hidden digits are carried in the internal workings of the machine.

  When it is necessary to add a percentage onto some amount, a short
cut is easily employed. For example: A new 145SR13 radial tyre costs
£18.29 plus VAT. What is the total cost? With VAT at 8% we wish to
find 8% of £18.29 and add this to the £18.29. The pedantic way is to
compute $18·29 \times ·08 + 18·29$ but what we really require is 8% of 18·29
plus 100% of 18·29, i.e. 108% of 18·29 so the simplified calculation is
$1·08 \times 18·29 = 19·7532$, so the cost is £19.75.

  Discounts are similarly handled. For example: In a record sale a
Bay City Rollers LP, priced at £2.73, is offered with '$12\frac{1}{2}\%$ off'. How
much will the LP cost?

  We require £2.73 less $12\frac{1}{2}\%$ of £2.73, i.e.

$$(100\% - 12\tfrac{1}{2}\%) \text{ of } £2.73$$

The calculation needed is

$$·875 \times 2·73 = 2·38875$$

so the cost is £2.39.

Some things to do

Exercise 4.2   Find the percentage composition by mass of a substance
  analysed to consist of 10·35 g nitrogen and 21·56 g oxygen only.

Exercise 4.3   Van der Waal's equation of state for n moles of a gas is

$$P = \frac{nRT}{V - nb} - \frac{n^2 a}{V^2}$$

where P is the pressure in pascals
V is the volume in m^3
T is the temperature in K
a and b are constants for each gas
R is the gas constant, 8.314 J mol^{-1} K^{-1}.

Find the pressure of three moles of CO_2 held at 400 K in a 5 litre cylinder, given that

$$a = 0.3636 \text{ Pa m}^6 \text{ mol}^{-2}$$

$$b = 4.267 \times 10^{-5} \text{ m}^3 \text{ mol}^{-1}$$

Exercise 4.4   Metabolic rate is roughly proportional to surface area per unit mass. For man it is about 260 cm^2 kg^{-1} and for some bacilli (cylindrical bacteria) about 5×10^7 cm^2 kg^{-1}. Show that these bacteria have a metabolic rate about 200000 times that of man. Assuming that man produces a new generation after 25 years how frequently might these bacteria reproduce?

Exercise 4.5   Micrococci are spherical bacteria. A micrococcus of diameter 200 μm consists mostly of water. Find the approximate mass of one micrococcus. The sulphur content is 0.01% by mass. Given that the mass of an atom of sulphur is 5.32×10^{-26} kg how many sulphur atoms will a micrococcus contain?

POWERS AND ROOTS

Most simple calculators have an $\boxed{x^2}$ key and this can be put to very good use if there is no general $\boxed{y^x}$ power key. For example if we require 2.9^{10}, then this is seen to be

$$2.9^8 \times 2.9^2 = ([(2.9)^2]^2)^2 \times 2.9^2$$

$$= 42070.723 \ldots \ldots$$

which is evaluated as follows:

Algebraic: 2.9 $\boxed{x^2}$ $\boxed{x^2}$ $\boxed{x^2}$ $\boxed{\times}$ 2.9 $\boxed{x^2}$ $\boxed{=}$

Reverse Polish: 2.9 $\boxed{x^2}$ $\boxed{\text{enter}}$ $\boxed{x^2}$ $\boxed{x^2}$ $\boxed{\times}$

with a minimum of entries.
   Another useful trick is illustrated as follows:

   'Find 1.72^{15}'.   This is $1.72^{16} \div 1.72$

so we simply enter 1.72, press $\boxed{x^2}$ four times to give $[([(1.72)^2]^2)^2]^2 = 1.72^{16}$ and then divide by 1.72, giving $1.72^{15} = 3411.3533 \ldots \ldots$

We see that any integer power can be quite easily handled. For negative integer powers we require the *reciprocal* of the number raised to the positive power, e.g.:

$$3.71^{-4} = \frac{1}{3 \cdot 71^4} = \left(\frac{1}{3 \cdot 71}\right)^4$$

and there are various methods of getting to the answer (0·0052784251), for example:

(i) 3·71 $\boxed{x^2}$ $\boxed{x^2}$ $\boxed{1/x}$

(ii) 3·71 $\boxed{1/x}$ $\boxed{x^2}$ $\boxed{x^2}$

(iii) Algebraic: 1 $\boxed{\div}$ 3·71 $\boxed{x^2}$ $\boxed{x^2}$ $\boxed{=}$

Reverse Polish: 1 $\boxed{\text{enter}}$ 3·71 $\boxed{\div}$ $\boxed{x^2}$ $\boxed{x^2}$

To find the perimeter of an ellipse with semi-axes $a = 4 \cdot 3$ and $b = 3 \cdot 1$ we use the approximate formula

$$P = 2\pi \sqrt{\left(\frac{a^2 + b^2}{2}\right)}$$

The key-stroke sequence might be

Algebraic:

4·3 $\boxed{x^2}$ $\boxed{+}$ 3·1 $\boxed{x^2}$ $\boxed{\div}$ 2 $\boxed{=}$ $\boxed{\sqrt{x}}$ $\boxed{\times}$ 2 $\boxed{\times}$ $\boxed{\pi}$ $\boxed{=}$

Reverse Polish:

4·3 $\boxed{x^2}$ $\boxed{\text{enter}}$ 3·1 $\boxed{x^2}$ $\boxed{+}$ 2 $\boxed{\div}$ $\boxed{\sqrt{x}}$ 2 $\boxed{\times}$ $\boxed{\pi}$ $\boxed{\times}$

giving the answer as 23·551471.

Non-integer powers, which include all the roots, are not easily handled without recourse to the $\boxed{y^x}$ function. Apart from the simplest cases, which may be dealt with by trial and error, some organized iterative scheme is needed.

The examples below indicate the kind of procedures which can be used. If the reader prefers he may go straight to Exercise 4.11 (page 50) skipping these procedures which do not require $\boxed{y^x}$.

(i) *Trial and error*

'Estimate the cube root of 71'. Let us represent $\sqrt[3]{71}$ by α. By inspection $4^3 < 71 < 5^3$ so $4 < \alpha < 5$ and we guess $\alpha \approx 4 \cdot 1$. We try

$4 \cdot 1^3 = 68 \cdot 9 \ldots \ldots$ and since this is a little too small we next try

$4 \cdot 2^3 = 74 \cdot 08 \ldots \ldots$ which is too large,

and so we must guess somewhere in between. A sensible choice might be 4·15.

$4\cdot15^3 = 71\cdot47$ .... too large so we try $4\cdot14$

$4\cdot14^3 = 70\cdot95$ .... and so $4\cdot14 < \alpha < 4\cdot15$ ...... and so on.

Something to do

Exercise 4.6   Use a trial-and-error method to estimate

  (*a*) $\sqrt{89}$ (*b*) $\sqrt[3]{123}$ (*c*) $\sqrt[4]{777777}$

(ii) *Organized iterative method*

'Estimate the square root of 44'. Here we wish to solve $x^2 = 44$ or $x = 44/x$.

If we make a guess, x_0, (say $x_0 = 6$) which is too small then the value of $44/x_0$ is bound to be too large (because we are dividing by too small a number) and so a generally better estimate is the average of these, i.e.

$$x_1 = \tfrac{1}{2}\left(x_0 + \frac{44}{x_0}\right).$$

We can devise a very useful recurrence relation

$$x_{r+1} = \tfrac{1}{2}\left(x_r + \frac{44}{x_r}\right), \quad r = 0, 1, 2, \ldots$$

and keep feeding in x_r on the right to produce successive estimates, x_{r+1}, which converge to the correct value.

Example:

Let $x_0 = 6$

then $x_1 = \tfrac{1}{2}\left(6 + \dfrac{44}{6}\right) = 6\cdot6666667$

$x_2 = \tfrac{1}{2}\left(6\cdot6666667 + \dfrac{44}{6\cdot6666667}\right) = 6\cdot6333333$

$x_3 = \tfrac{1}{2}\left(6\cdot6333333 + \dfrac{44}{6\cdot6333333}\right) = 6\cdot6332496$

$x_4 = \tfrac{1}{2}\left(6\cdot6332496 + \dfrac{44}{6\cdot6332496}\right) = 6\cdot6332496$

and we have the answer correct to 8 s.f.

Something to do

Exercise 4.7   Find $\sqrt{7}$ and $\sqrt{77}$ by an iterative procedure and compare with the directly calculated values. Devise procedures for use with your calculator to minimize writing down intermediate results and to minimize key strokes.

A note to the reader: The next two sections, (iii) and (iv), are rather harder to follow. You may prefer to go straight to Exercise 4.11 (p. 50).

(iii) *Finding all roots using only the square root function—some simpler cases*

If we require $x^{\frac{1}{4}}$, i.e. the fourth root, then it can be found using $\boxed{\sqrt{x}}$ $\boxed{\sqrt{x}}$. Similarly $x^{\frac{1}{8}}$ just needs $\boxed{\sqrt{x}}$ $\boxed{\sqrt{x}}$ $\boxed{\sqrt{x}}$. Clearly we can find

$$x^{\frac{1}{2}}, \ x^{\frac{1}{4}}, \ x^{\frac{1}{8}}, \ x^{\frac{1}{16}}, \ \ldots .$$

Are others possible? Multiplying $x^{\frac{1}{2}}$ and $x^{\frac{1}{4}}$ will give $x^{\frac{3}{4}}$, and so a whole host of other roots can be found, all those with 2, 4, 8, ... in the denominator of the fractional power. But can we get $x^{\frac{1}{3}}$ or $x^{\frac{1}{5}}$ etc.? We can if we can express the power as a sum of the known roots $\frac{1}{2}, \frac{1}{4}, \frac{1}{8}, \ldots$. Writing $\frac{1}{3}$, for example, as a binary decimal we have $\frac{1}{3} = \cdot01010101 \ldots$ (base 2). This means that

$$\frac{1}{3} = \frac{1}{4} + \frac{1}{16} + \frac{1}{64} + \frac{1}{256} + \frac{1}{1024} + \cdots .$$

Each of $x^{\frac{1}{4}}, x^{\frac{1}{16}}, \ldots$ can be evaluated just using $\boxed{\sqrt{x}}$, so by multiplying together the first few of these we will get an approximation to $x^{\frac{1}{3}}$. The powers of x not taken (perhaps $x^{1/4096}, \ldots$) are close to 1 (slightly above) and so this method gives an underestimate for the root.

Using an 8-digit calculator without $\boxed{y^x}$ but with $\boxed{\sqrt{x}}$ and one memory (Busicom 811DBR) Table 4.1 was produced in estimating $\sqrt[3]{10}$:

TABLE 4.1   *Estimating $\sqrt[3]{10}$ without using* $\boxed{y^x}$

Power of next term	Next term	Binary form of total power	Total power	Decimal form of total power	Root estimate
1/4	1·7782793	·01	$\frac{1}{4}$	·25	1·7782793
1/16	1·1547819	·0101	$\frac{1}{4}+\frac{1}{16}$	·3125	2·0535247
1/64	1·0366328	·010101	$\frac{1}{4}+\frac{1}{16}+\frac{1}{64}$	·328125	2·1287510
1/256	1·0090349	etc.	etc.	·33203125	2·1479840
1/1024	1·0022510	. . .	. . .	·33300781	2·1528191
1/4096	1·0005622	. . .	. . .	·33325195	2·1540294
1/16384	1·0001404	. . .	. . .	·33331299	2·1543318
1/65536	1·0000350	. . .	. . .	·33332825	2·1544072
1/262144	1·0000086	. . .	. . .	·33333206	2·1544257
1/1048576	1·0000020	. . .	. . .	·33333302	2·1544300
1/4194304	1·0000004	. . .	. . .	·33333325	2·1544308
1/16777216	1·	. . .	. . .	·33333331	2·1544308

This estimate for $\sqrt[3]{10}$ is correct to six s.f.

Calculators carrying more digits may in fact give the result to machine accuracy, i.e. exactly the same as using $\boxed{\sqrt[x]{y}}$ or $\boxed{y^x}$.

The key-stroke sequence for $\sqrt[x]{}$ is very simple and only one memory is needed (no doubt Reverse Polish procedures only using the stack can be devised):

	Algebraic key strokes	Display	Binary decimal part	Reverse Polish key strokes
Initial set-up	10	10·		10
	$\boxed{\sqrt{x}}$ $\boxed{\sqrt{x}}$			$\boxed{\sqrt{x}}$ $\boxed{\sqrt{x}}$
	$\boxed{STO}$ $\boxed{\times}$	1·7782794	·01	$\boxed{STO}$
Cycle	$\boxed{RCL}$ $\boxed{\sqrt{x}}$			$\boxed{RCL}$ $\boxed{\sqrt{x}}$
	$\boxed{\sqrt{x}}$ $\boxed{STO}$			$\boxed{\sqrt{x}}$ $\boxed{STO}$
	$\boxed{\times}$	2·053525	01	$\boxed{\times}$
Cycle	$\boxed{RCL}$ $\boxed{\sqrt{x}}$			$\boxed{RCL}$ $\boxed{\sqrt{x}}$
	$\boxed{\sqrt{x}}$ $\boxed{STO}$			$\boxed{\sqrt{x}}$ $\boxed{STO}$
	$\boxed{\times}$	2·1287517	01	$\boxed{\times}$
and so on		. . . . .	. .	and so on

Something to do

Exercise 4.8   Using only the $\boxed{\sqrt{x}}$ function find

(a) $\sqrt[7]{12345}$ $[\frac{1}{7} = \dot{0}\cdot 0\dot{1}$ (base 2)]

(b) $\sqrt[5]{99}$ $[\frac{1}{5} = \cdot \dot{0}01\dot{1}$ (base 2)]

(iv) *Finding roots for which the binary decimal is more complicated*

To find $\sqrt[13]{n}$ is harder because the binary decimal does not repeat so soon or so neatly:

$$\frac{1}{13} = \cdot\dot{0}0010011101\dot{1} \text{ (base 2)}$$

A much longer key-stroke cycle is needed but it need be repeated many fewer times. One workable scheme is:

	Algebraic key strokes	Binary decimal part	Reverse Polish key strokes
Initial set-up	n		n
	$\boxed{\sqrt{x}}$ $\boxed{\sqrt{x}}$		$\boxed{\sqrt{x}}$ $\boxed{\sqrt{x}}$
	$\boxed{\sqrt{x}}$ $\boxed{\sqrt{x}}$		$\boxed{\sqrt{x}}$ $\boxed{\sqrt{x}}$
	$\boxed{STO}$ $\boxed{\times}$	·0001	$\boxed{STO}$

(*continued*)

	Algebraic key strokes	Binary decimal part	Reverse Polish key strokes

Key stroke cycle

RCL √x √x √x

STO ×   001

RCL √x STO ×   1

RCL √x STO ×   1

RCL √x √x STO

×   01

RCL √x STO ×   1

RCL √x √x √x

√x STO ×   0001

(keystroke cycle as for algebraic)

Note that there is a √x for each binary digit and a × for each 1.

Some things to do

Exercise 4.9   Use the above procedure to find $\sqrt[13]{987654}$.

Exercise 4.10   Devise suitable procedures using M× to find various roots, and if possible test them.

An interesting article by Wynne Wilson (1976) shows how √x can be used to generate many scientific functions.

Some things to do (*General*)

Exercise 4.11   Virtually all metallic elements adopt one or more of three crystalline structures: cubic close-packed, hexagonal close-packed, body-centred cubic. The inter-atomic distances are given by the formulae:

$$d = \sqrt[3]{\left\{\frac{A \times \sqrt{2}}{N_A \times \rho}\right\}} \quad \text{for cubic close-packed and hexagonal close-packed}$$

$$d = \sqrt[3]{\left\{\frac{A \times 3\sqrt{3}}{N_A \times 4\rho}\right\}} \quad \text{for body-centred cubic}$$

where N_A is the Avogadro Constant ($\approx 6{\cdot}023 \times 10^{23}$ mol^{-1})

A is the mass of one mole in kg

ρ is the density (kg m^{-3})

d is the inter-atomic distance (m)

Compute the inter-atomic distances for silver (cubic close-packed, $A = 0.1079$, $\rho = 10.5 \times 10^3$ and for potassium (body-centred cubic, $A = 0.0391$, $\rho = 0.87 \times 10^3$).

Exercise 4.12 Two pulleys of radii 20 cm and 12 cm have their centres 1.95 m apart. What length of belt is required to pass round them? An approximate formula is: $L = \pi R_1 + \pi R_2 + 2\sqrt{(d^2 - (R_1 - R_2)^2)}$ where R_1, R_2 are the radii and d is the distance between the centres of the pulleys.

Exercise 4.13 Three resistors of 290 Ω, 325 Ω and 440 Ω are connected in parallel. Find the equivalent resistance, R, given that

$$\frac{1}{R} = \frac{1}{R_1} + \frac{1}{R_2} + \frac{1}{R_3}.$$

Exercise 4.14 The flight Mach number, M, of an aircraft can be found from the Calibrated Air Speed reading and Pressure Altitude reading. The formula is

$$M = \sqrt{\left(\frac{2}{\gamma-1}\right)\left\{\left(\left[\left\{\left(1+\frac{\gamma-1}{2}\left[\frac{CAS}{a_{SL}}\right]^2\right)^{\gamma/(\gamma-1)}-1\right\}\right.\right.\right.}$$
$$\left.\left.\left.\times[1-(6.875\times10^{-6})H]^{-5.266}\right]+1\right)^{(\gamma-1)/\gamma}-1\right\}\right)$$

where CAS is the Calibrated Air Speed (knots)

H is the Pressure Altitude reading (feet)

a_{SL} is the Speed of Sound at Sea Level at 15°C at 1 atmosphere air pressure (knots)

γ is c_p/c_v—the ratio of the specific heats of air at constant pressure and constant volume

Given that $a_{SL} = 661.5$ knots

and $\gamma = 1.40$

find the flight Mach number of an aircraft when the instruments show a Pressure Altitude of 22300 feet at a CAS of 385 knots.

Exercise 4.15 Dreyer found an empirical formula relating mass in grams (M) and sitting height in cm (h) for adults:

$$M^{0.319} = 0.3803\ h$$

What should the sitting height be for a person of mass 75200 g? Check the formula empirically yourself.

Exercise 4.16 (a) Multi-stage rockets are more efficient for attaining high velocities than are single stage rockets and it is best if the stages are unequal in size. D. N. Burghes (1974) showed that for a rocket with

n stages of masses m_1, m_2, ... m_n the optimum choices of stage masses are:

$$m_r = (1+\beta)^{1-r/n} \times \beta^{(r-1)/n} \times [(1+\beta)^{1/n} - \beta^{1/n}] \times M$$

$$\text{for } r = 1, 2, \ldots n$$

where $M = m_1 + m_2 + \ldots + m_n$

　　　P is the mass of the payload (e.g. a satellite to be put into orbit)
　　　$\beta = P/M$

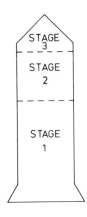

Fig. 4.1

Find the optimum masses for an n-stage rocket for which $\beta = 0{\cdot}01$ and $M = 10000$ kg where

(i) $n=2$　　　(ii) $n=3$　　　(iii) $n=4$　　　(iv) $n=5$

(b) The maximum final velocity attainable by an n-stage rocket is given by

$$v_f = -nc \ln [1 - \varepsilon + \varepsilon\beta^{1/n}/(1+\beta)^{1/n}]$$

where each stage contributes an equal increment, v_f/n, to the final velocity. c is the velocity of the escaping exhaust gas, relative to the rocket. ε is the fraction of the initial mass of the rocket which consists of fuel ($\varepsilon = $ fuel mass$/M$).

　　For the typical values $\varepsilon = 0{\cdot}8$, $c = 2{\cdot}5$ km s^{-1}, $\beta = 0{\cdot}01$ calculate v_f for n-stage rockets where

(i) $n=2$　　　(ii) $n=3$　　　(iii) $n=4$　　　(iv) $n=5$

How many stages do you think interplanetary rockets should have? Is there a limit to the final velocity attainable?

SEQUENCES AND SERIES

A sequence is a set of numbers. For example: 1, 2, 3, 4, 5 is a finite sequence and 1, 2, 3, 4, 5, ... is an infinite sequence. A series is the summation of a set of numbers, for example $\frac{1}{2}+\frac{1}{4}+\frac{1}{8}+\frac{1}{16}$ is a finite series and $\frac{1}{2}+\frac{1}{4}+\frac{1}{8}+\frac{1}{16}+$ ... is an infinite series. Generally there will be a simple rule indicating how the next term is to be calculated given all the previous terms.

Arithmetic progressions

The sequence of counting numbers (the 'natural numbers') 1, 2, 3, 4, ... is a particularly simple sequence. Each term is one more than the last and the first number is 1. Any sequence for which the next term is produced by adding a fixed number (positive, zero, or negative) to the previous term is called an Arithmetic Progression (A.P.).

In general an A.P. is a sequence

$$a, a+d, a+2d, a+3d, a+4d, \ldots$$

where a is the first term and d is the common difference.

Examples: 6,   9,   12,   15,   18, ... $(a=6, d=3)$
          20, $16\frac{1}{2}$,   13, $9\frac{1}{2}$, ... $(a=20, d=-3\frac{1}{2})$

The terms of an A.P. are very easily generated using a calculator. The common difference is stored in the memory (or one's head!) and the first term, a, is entered into the machine and d repeatedly added to it.

Example. Calculate the successive terms of the A.P. defined by $a=11$, $d=7$.

Algebraic key strokes	Display	Comment
7 STO	7·	d
+ 11 +	18·	$a+d$
RCL +	25·	$a+2d$
RCL +	32·	$a+3d$
RCL +	39·	$a+4d$
and so on	...	....

The Reverse Polish procedure could make use of a memory, much as does the above algebraic procedure, or use can be made of the stack, provided that the machine is designed so that when the stack drops the previous contents of the T register (now moved down to the Z register) are reproduced in the T register. If this is so (it is on Hewlett Packard machines and on the Corvus 500 and may be universally so) then the stack can be used to hold the common difference, d.

Reverse Polish key strokes (using stack)	*Display*	*Comment*	*Reverse Polish key strokes (using memory)*
7 [enter] [enter] [enter]	7·	d	7 [STO] [enter]
11 [+]	18·	$a+d$	11 [+]
[+]	25·	$a+2d$	[RCL] [+]
[+]	32·	$a+3d$	[RCL] [+]
[+]	39·	$a+4d$	[RCL] [+]
and so on	. . .	. . . .	and so on

If a particular term of the A.P. is required, the nth term say, then the formula $a+(n-1)d$ is used. The best procedure here is to enter $n-1$ first, then multiply by d, then add a.

If the summation of n terms of an A.P. is required, i.e.

$$S_n = a + \overline{a+d} + \overline{a+2d} + \overline{a+3d} + \ldots + \overline{a+(n-1)d}$$

the appropriate formulae are

$$S_n = \frac{n}{2}\{2a+(n-1)d\} \quad \text{or} \quad S_n = \left(\frac{l-a+d}{d}\right)\left(\frac{a+l}{2}\right) \quad \text{or} \quad S_n = \frac{n}{2}\{a+l\}$$

where l is the last term, i.e. $a+(n-1)d$

Example:

Find $1+4+7+10+13+16+19$.

Here $n=7$, $a=1$, $d=3$, $l=19$, so

$$S_n = \frac{n}{2}(2a+(n-1)d) = \frac{7}{2}(2+6\times3) = 70$$

or

$$S_n = \frac{7}{2}(1+19) = 70$$

Something to do

Exercise 4.17   Sum the series

   (a) $6+7\cdot5+9+10\cdot5+ \ldots +36$

   (b) $42+37+32+ \ldots (23 \text{ terms})$

   (c) $44+41+38+ \ldots +11$ (12 terms)

Geometric Progressions

The sequence of powers of 2: 1, 2, 4, 8, 16, . . . is a simple example of a Geometric Progression (G.P.). Each term is twice the previous term, and the first term is 1. Any sequence for which the next term is a constant multiple of the previous term is a Geometric Progression.

In general a G.P. is a sequence

$$a, ar, ar^2, ar^3, \ldots$$

where a is the first term and r is the common ratio.

Examples:

2, 6, 18, 54, 162, . . . $(a=2, r=3)$

200, -100, 50, -25, 12·5, . . . $(a=200, r=-\frac{1}{2})$

The terms of a G.P. are very easily generated using a calculator—r is stored (or remembered), a is entered into the machine, and repeatedly multiplied by r.

Example. Calculate the successive terms of the G.P. defined by $a=4\cdot3$, $r=-0\cdot9$.

Algebraic key strokes	Display	Comment
·9 [+/−] [STO]	− ·9	r
[×] 4·3 [×]	− 3·87	ar
[RCL] [×]	3·483	ar^2
[RCL] [×]	− 3·1347	ar^3
and so on	. . .	. . .

Reverse Polish key strokes (using stack)	Display	Comment	Reverse Polish key strokes (using memory)
·9 [+/−] [enter]			·9 [+/−]
[enter] [enter]	− ·9	r	[STO]
4·3 [×]	− 3·87	ar	4·3 [×]
[×]	3·483	ar^2	RCL [×]
[×]	− 3·1347	ar^3	RCL [×]
and so on	. . .	. . .	and so on

The nth term of a G.P. is ar^{n-1} which is best computed using the [y^x] function if available but which can of course be found by repeated multiplication and appropriate use of the [x^2] key.

If the summation of n terms of a G.P. is wanted, i.e.

$$S_n = a + ar + ar^2 + ar^3 + \ldots + ar^{n-1}$$

the appropriate formula is

$$S_n = a\,\frac{(1-r^n)}{1-r} = a\,\frac{(r^n-1)}{r-1}$$

Example:

Find $4 + 8 + \ldots + 128$.

Here $a = 4$, $r = 2$, $n = 6$, so

$$S_n = 4\,\frac{(2^6-1)}{2-1} = 4(64-1) = 252$$

Providing r lies within the limits -1 to $+1$ (i.e. $-1 < r < +1$) it is possible to sum an infinite G.P.

In this case the previous formula $S_n = a(1-r^n)/(1-r)$ is simplified by the fact that $r^n \to 0$ as $n \to \infty$. (The symbol $\to$ means 'tends to' and ∞ is the infinity sign denoting a number large without bound.) Thus

$$S_\infty = \frac{a}{1-r}$$

Example:

Find $1 - \dfrac{1}{3} + \dfrac{1}{9} - \dfrac{1}{27} + \dfrac{1}{81} - \dfrac{1}{243} + \ldots$.

Here $a = 1$, $r = -\frac{1}{3}$, so

$$S_\infty = \frac{1}{1+\frac{1}{3}} = \tfrac{3}{4}$$

It is interesting to compute the partial sums of an infinite G.P. (for $-1 < r < +1$) to observe how the series tends towards its limit. For example:

Term added	Partial sum (*to 3 d.p.*)
1	$1 \cdot 000$
$-\dfrac{1}{3}$	$0 \cdot 667$
$+\dfrac{1}{9}$	$0 \cdot 778$
$-\dfrac{1}{27}$	$0 \cdot 741$
$+\dfrac{1}{81}$	$0 \cdot 753$
$-\dfrac{1}{243}$	$0 \cdot 749$
$+\dfrac{1}{729}$	$0 \cdot 750$

Such calculations can of course be aided by use of the calculator. To perform the whole set of computations on an algebraic logic machine with no intermediate numerical entries to be necessary requires two memories (one for a and one for r), or one memory and the stack on a Reverse Polish logic machine.

Some things to do

Exercise 4.18   Statistical evidence suggests that for many animals, and particularly for birds, mortality rate is constant with age (ignoring the very high death rate for the very young), at about a one-in-three chance per year. This means that the numbers of birds of different ages (1 year old, 2 years old, ...) should form a sequence approximating to a geometric progression with common ratio $r = 1 - m$, where m is the fraction of mortalities in a year. For lapwings (Vanellus vanellus) and wood pigeons (Columba palumbus) the following figures were given by J. B. S. Haldane (1953) and R. K. Murton (1966):

Lapwings $m = 0 \cdot 342$		*Wood pigeons* $m = 0 \cdot 356$	
Age (years)	*Population*	*Age* (years)	*Population*
1	198	1	204
2	134	2	123
3	90	3	77
4	51	4	52
5	48	5	34
6	23	6	27
7	21	7	22
8	9	8	14
9	6	9	8
10	5	10	6
11	6	11	3
12	1	12	1
13	0	13	1
14	1	14	1

Taking the annual mortality rate for lapwings as $34 \cdot 2\%$ (i.e. $r = 1 - 0 \cdot 342 = 0 \cdot 658$) and for wood pigeons $35 \cdot 6\%$ ($r = 0 \cdot 644$), generate two geometric series, one for $a = 198$, $r = 0 \cdot 658$ the other for $a = 204$, $r = 0 \cdot 644$ and see how well the experimental data fit the theoretical figures and support the hypothesis of constant mortality rate.

Exercise 4.19   The collared dove (*Streptopelia decaocto*) has been invading Western Europe (since the 1930's) and has been rapidly spreading throughout Britain since it first arrived in 1952 R. Hudson

(1965) presented the following table to show the 'geometric population increase' for the period 1955–1964:

	1955	1956	1957	1958	1959
S. England	4	16	35	80	130
N. England	—	—	5	10	40
Wales	—	—	—	—	—
Scotland	—	—	5	10	25
Ireland	—	—	—	—	10
Totals	4	16	45	100	205
	1960	1961	1962	1963	1964
S. England	490	1,270	3,590	7,810	14,545
N. England	100	400	720	1,610	2,565
Wales	—	20	45	95	270
Scotland	70	185	260	580	1,115
Ireland	15	25	35	105	360
Totals	675	1,900	4,650	10,200	18,855

Test the data to see how closely they follow a geometric progression. (Try $r = 2\cdot3$ for various starting values, a.)

Exercise 4.20   A simple exhaust pump (see fig. 4.2) can be used to evacuate the gas from the large vessel (volume V). With one stroke

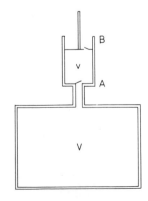

Fig. 4.2

the pressure is reduced by the constant factor $V/(V+v)$ so with n depressions of the pump the pressure p_n is (approximately) given by

$$p_n = p_0 \times \left\{ \frac{V}{V+v} \right\}^n,$$

where p_0 is the initial pressure.

For $p_0 = 50000$ Pa, $V = 6000$ cm^3, $v = 90$ cm^3 calculate the successive pressures $p_1, p_2, \ldots p_{20}$.

How many strokes of the piston would be needed to reduce the pressure to below 100 Pa?

Exercise 4.21 Many series involve π ($= 3\cdot14\ldots$) and some are suitable for finding the value of π accurately.

For example

$$\frac{\pi^8}{9450} = \frac{1}{1^8} + \frac{1}{2^8} + \frac{1}{3^8} + \frac{1}{4^8} + \frac{1}{5^8} + \ldots$$

Use the following procedure to find π accurately, and compare with the value shown on pressing the π key.

(i) Compute

$$1 + \frac{1}{2^8} + \frac{1}{3^8} + \ldots$$

adding terms until they are too small to affect the total. You can use n $\boxed{x^2}$ $\boxed{x^2}$ $\boxed{x^2}$ $\boxed{1/x}$ to give $1/n^8$ and so can do all the calculations directly in the display without using a memory or parentheses.

(ii) Multiply by 9450 to give π^8.

(iii) Take the 8th root using $\boxed{\sqrt[x]{y}}$ 8 or $\boxed{y^x}$ $\cdot125$ or $\boxed{\sqrt{x}}$ $\boxed{\sqrt{x}}$ $\boxed{\sqrt{x}}$.

RECURRENCE RELATIONS

If the terms of a sequence are $u_1, u_2, u_3, u_4, u_5, \ldots$ where the u_i are any numbers, then it is usually possible to write down an expression for a general term u_n relating it to one or more previous terms.

For an A.P. we can write the recurrence relation

$$u_n = u_{n-1} + d \quad (\text{with } u_1 = a)$$

$\uparrow$ new term $\quad$ $\uparrow$ previous term $\quad$ $\uparrow$ common difference

and u_n depends on the previous term u_{n-1}. Similarly for a G.P. we have the recurrence relation

$$u_n = r \times u_{n-1} \quad \text{(with } u_1 = a)$$

↑   ↑   ↑
new   common   previous
term   ratio   term

A rather different recurrence relation is

$$u_n = 2u_{n-1} + u_{n-2} \quad \text{(with } u_1 = 1, \ u_2 = 4)$$

which actually involves the previous *two* terms to define the next term of the sequence

$u_1 = 1$ (given)

$u_2 = 4$ (given)

$u_3 = 2u_2 + u_1$ (putting $n = 3$ in the recurrence relation)

     $= 2 \times 4 + 1 = 9$

similarly $u_4 = 2u_3 + u_2 = 22$

                       $u_5 = 2u_4 + u_3 = 53$

and so on.

Clearly the calculator is useful for generating such sequences and the next section will consider such an application for a sequence of particular mathematical importance.

THE FIBONACCI SEQUENCE

The thirteenth century textbook *Liber Abaci* of Leonardo of Pisa included the following hypothetical problem in population dynamics:

'What is the number of pairs of rabbits at the beginning of each month of a particular year if a single pair of newly born rabbits is put into a walled enclosure on January 1st? Each pair breeds a new pair at the beginning of the second month after birth and an additional pair each month thereafter. It is to be assumed that no deaths occur during the year.'

It is fairly easy to analyse this problem:

Beginning of month	Pairs	Comment
1	1	The original newly born pair.
2	1	The original pair 1 month old.
3	2	Original pair + new offspring.
4	3	One more pair from original pair.
5	5	The 3 pairs from month 4 plus 2 more pairs from those old enough to breed again (i.e. from those alive 2 months ago).

and so on.

We see that for month n the total rabbit population u_n, is given by

$$u_n = u_{n-1} + u_{n-2}$$

↑	↑
All those pairs alive last month and so still alive	One more pair from each of those pairs alive 2 months ago

Also $u_1 = u_2 = 1$.

Thus we have a recurrence relation to define the Fibonacci sequence 1, 1, 2, 3, 5, 8, 13, 21, 34, 55, ... which has many fascinating properties. Many flowering plants tend to have numbers of petals which are members of the Fibonacci sequence (something to investigate!) and the occurrence of stems on plants and branches on trees often demonstrate this numerical pattern (see F. W. Land, 1975).

The calculator provides a simple method of generating new terms. If no memory is used it is necessary at each stage to add in the previous term to the present term (already in the display) in order to calculate the next term:

Algebraic key strokes	Display	Comment	Reverse Polish key strokes
1 [+]	1·	u_2	1 [enter]
1 [+]	2·	u_3	1 [+]
1 [+]	3·	(adds u_2 to display) u_4	1 [+]
2 [+]	5·	(adds u_3 to display) u_5	2 [+]
3 [+]	8·	(adds u_4 to display) u_6	3 [+]

What if we have one memory available? Surprisingly, the task is not appreciably easier. At first it would seem that we should be able to use the memory to store the previous term (u_{n-2}) and add it to the present term in the display (u_{n-1}) to generate the new term (u_n). However, doing this destroys the u_{n-1} term which we should now allocate to the memory!

Algebraic key strokes	Display	Memory	Reverse Polish key strokes
1	1· [u_1]	—	1 [enter]
[STO] [+]	1· [u_2]	1· [u_1]	[STO]
[RCL] [+]	2· [u_3]	1· [u_2]	[RCL] [+]
[RCL] [+]	3· [u_4]	1· [u_2]	[RCL] [+]

and now we have lost the 2· and cannot recover it without destroying the 3· which is wanted for calculations of the next term. The problem is basically that of wishing to interchange the contents of the display and the memory without having some intermediate store available.

With Reverse Polish machines the rolling stack can be made use of and one memory will then suffice:

	Reverse Polish key strokes	Display	Comment
Initial setting up	1 [enter]	1·	u_1, u_2
	2 [enter] [STO]	2·	u_3
cycle 1	[R↓] [R↓]	1·	$[u_2]$ recalled from stack
	[RCL] [+] [STO]	3·	u_4
cycle 2	[R↓] [R↓]	2·	$[u_3]$ recalled from stack
	[RCL] [+] [STO]	5·	u_5
cycle 3	[R↓] [R↓]	3·	$[u_4]$ recalled from stack
	[RCL] [+] [STO]	8·	u_6

etc.

If an algebraic logic calculator being used has the interchange key [X⇌M] then all is well, and the correct procedure is

	Algebraic key strokes	Display	Memory
Initial setting up	1	1· u_1	—
	[STO]	1· u_2	1· $[u_1]$
	[+] [RCL] [=]	2· u_3	1· $[u_2]$
cycle 1	[X⇌M]	1· u_2	2· $[u_3]$
	[+] [RCL] [=]	3· u_4	2· $[u_3]$
cycle 2	[X⇌M]	2· u_3	3· $[u_4]$
	[+] [RCL] [=]	5· u_5	3· $[u_4]$

etc.

Thus the Fibonacci sequence can be very easily generated with a repeated four key-stroke sequence on an algebraic machine with one memory and the [X⇌M] key.

Some calculators do not have [X⇌M] but have another useful key [M+] or [Σ] which adds the contents of the display directly into the memory. A procedure for generating the Fibonacci numbers using [M+] and a single memory is as follows:

	Algebraic key strokes	Display	Memory	Reverse Polish key strokes	
Initial setting up	1	$1 \cdot [u_1]$	—	1 [enter]	Initial setting up
	[STO]	$1 \cdot$	$1 \cdot$	[STO]	
Cycle 1	[M+]	$1 \cdot$	$2 \cdot$	[Σ]	Cycle 1
	[−] [RCL] [=]	$-1 \cdot$	$2 \cdot$	[RCL] [−]	
	[+/−]	$1 \cdot [u_2]$	$2 \cdot$	[+/−]	
Cycle 2	[M+]	$1 \cdot$	$3 \cdot$	[Σ]	Cycle 2
	[−] [RCL] [=]	$-2 \cdot$	$3 \cdot$	[RCL] [−]	
	[+/−]	$2 \cdot [u_3]$	$3 \cdot$	[+/−]	

etc.   etc.

Here a five key-stroke sequence is needed for algebraic machines and the procedure itself is a little less straightforward.

Note: The Reverse Polish key-stroke sequence using [Σ] may not work as shown. On some machines (e.g. HP 45) when [Σ] is pressed the number in the display disappears and a counter is shown, indicating how many [Σ] entries there have been. The required number can be recovered by pressing the [last x] function key immediately after [Σ] has been pressed.

Some things to do

Exercise 4.22   (*a*) Generate the Fibonacci-like sequence

$$u_n = u_{n-1} + u_{n-2}; \quad u_1 = 3, u_2 = 2$$

(*b*) Generate the sequence defined by

$$u_n = u_{n-1} - u_{n-2} + u_{n-3}; \quad u_1 = 1, u_2 = 2, u_3 = 4$$

Exercise 4.23   Devise a suitable automatic procedure for generating the Fibonacci sequence if two memories are available.

Exercise 4.24   Check the following properties of the Fibonacci sequence by testing their validity for $n = 1$, $n = 2$, and so on.

$$(a) \ u_1 + u_2 + u_3 + \ldots + u_n = u_{n+2} - 1$$
$$(b) \ u_1 + u_3 + u_5 + \ldots + u_{2n-1} = u_{2n}$$
$$(c) \ u_1^2 + u_2^2 + u_3^2 + \ldots u_n^2 = u_n . u_{n+1}$$

Exercise 4.25   The pair of related sequences defined by

$$s_1 = \sqrt{2}; \quad s_{n+1} = \sqrt{(2 + s_n)}$$
$$t_1 = \sqrt{2}; \quad t_{n+1} = 2^{n+1} \sqrt{(2 - s_n)}$$

produce interesting limiting values as n is increased (more and more terms taken) and is discussed by M. A. B. Deakin (1974).

Show that $s_n \to 2$ and find the value to which t_n tends. (Devise a suitable scheme to use the calculator efficiently, using memories as appropriate.) Notice that accuracy in computing the t_{n+1} values falls away after a while.   Why?

THE FIBONACCI SEQUENCE AND BINET'S FORMULA

The generation of the terms of the Fibonacci sequence using the repeated application of the recurrence relation can be tedious, especially if one particular term, u_n, is sought rather than all the terms $u_1, u_2, \ldots u_n$. Can we, then, find u_{12} (say) directly? The answer is 'yes' because a formula for u_n has been derived. It is called the Binet formula after the French mathematician, Binet. The formula is

$$u_n = \frac{\left(\frac{1+\sqrt{5}}{2}\right)^n - \left(\frac{1-\sqrt{5}}{2}\right)^n}{\sqrt{5}}$$

Putting

$$\alpha = \frac{1+\sqrt{5}}{2} = 1 \cdot 61803 \ldots \quad \text{and} \quad \beta = \frac{1-\sqrt{5}}{2} = -0 \cdot 61803 \ldots$$

we get the possibly simpler-looking formula

$$u_n = \frac{\alpha^n - \beta^n}{\sqrt{5}}$$

When using a calculator to find a particular term, u_n, by the Binet formula it is important to remember that β $(=(1-\sqrt{5})/2)$ is negative and so care is needed in evaluating

$$\left(\frac{1-\sqrt{5}}{2}\right)^n$$

since it is impermissible to raise a negative number to a power using the $\boxed{y^x}$ function. So if n is *even* use the fact that

$$\left(\frac{1-\sqrt{5}}{2}\right)^n = \left(\frac{\sqrt{5}-1}{2}\right)^n$$

and if n is *odd* use

$$\left(\frac{1-\sqrt{5}}{2}\right)^n = -\left(\frac{\sqrt{5}-1}{2}\right)^n.$$

Example:
To find

$$u_{11} = \frac{\left(\frac{1+\sqrt{5}}{2}\right)^{11} - \left(\frac{1-\sqrt{5}}{2}\right)^{11}}{\sqrt{5}} = \frac{\left(\frac{1+\sqrt{5}}{2}\right)^{11} + \left(\frac{\sqrt{5}-1}{2}\right)^{11}}{\sqrt{5}}$$

Algebraic key strokes	Display	Comment	Reverse Polish key strokes
5 $\boxed{\sqrt{x}}$ $\boxed{+}$ 1			5 $\boxed{\sqrt{x}}$ 1 $\boxed{+}$
$\boxed{\div}$ 2 $\boxed{=}$	1·618034	α	2 $\boxed{\div}$
$\boxed{y^x}$ 11 $\boxed{=}$			11 $\boxed{y^x}$
$\boxed{STO}$	199·00503	α^{11}	$\boxed{STO}$
5 $\boxed{\sqrt{x}}$ $\boxed{-}$ 1			5 $\boxed{\sqrt{x}}$ 1 $\boxed{-}$
$\boxed{\div}$ 2 $\boxed{=}$	·61803399	$-\beta$	2 $\boxed{\div}$
$\boxed{y^x}$ 11 $\boxed{=}$	·005025	$-\beta^{11}$	11 $\boxed{y^x}$
$\boxed{+}$ $\boxed{RCL}$			$\boxed{RCL}$ $\boxed{+}$ 5
$\boxed{\div}$ 5 $\boxed{\sqrt{x}}$ $\boxed{=}$	89·	$(\alpha^{11} - \beta^{11})/\sqrt{5}$	$\boxed{\sqrt{x}}$ $\boxed{\div}$

Note: 'sum of products' logic machines (e.g. Texas Instruments) will need an $\boxed{=}$ before each $\boxed{\div}$.

In fact the key strokes can be considerably reduced when it is noticed that $\beta = -1/\alpha$:

Algebraic key strokes	Display	Comment	Reverse Polish key strokes
5 $\boxed{\sqrt{x}}$ $\boxed{+}$ 1			5 $\boxed{\sqrt{x}}$ 1 $\boxed{+}$
$\boxed{\div}$ 2 $\boxed{=}$	1·618034	α	2 $\boxed{\div}$
$\boxed{y^x}$ 11 $\boxed{=}$			11 $\boxed{y^x}$
$\boxed{STO}$	199·00503	α^{11}	$\boxed{STO}$
$\boxed{1/x}$	·005025	$-\beta^{11}$	$\boxed{1/x}$
$\boxed{+}$ $\boxed{RCL}$ $\boxed{\div}$			$\boxed{RCL}$ $\boxed{+}$ 5
5 $\boxed{\sqrt{x}}$ $\boxed{=}$	89·	u_{11} very accurately	$\boxed{\sqrt{x}}$ $\boxed{\div}$

It will be seen that the second term $\{-\beta^{11}/\sqrt{5}\}$ makes very little contribution. This is always so because $\beta = -0.61803\ldots$ has magnitude rather less than 1 and thus β^n for n a positive integer is always less than unity and gets smaller with increasing n, whereas α_n is always greater than unity and rapidly becomes very large with increasing n. This leads to a simpler formula. $\beta^n/\sqrt{5}$ is in fact numerically largest when $n=1$ in which case $\beta/\sqrt{5} = -0.2\ldots$ so the second term $\beta^n/\sqrt{5}$ is always numerically less than $\frac{1}{2}$, so the term $\alpha^n/\sqrt{5}$ always gives the correct value when rounded off to the nearest integer. The simpler formula is:

$$u_n \approx \frac{\alpha^n}{\sqrt{5}} \quad \text{or} \quad u_n \approx \frac{1}{\sqrt{5}} \left(\frac{1+\sqrt{5}}{2} \right)^n$$

e.g. Using this formula $u_6 \approx 8·025$ hence $u_6 = 8$.

This simplified formula provides a very neat way of generating the terms of the Fibonacci sequence by calculator:

(i) Calculate $\alpha = (1 + \sqrt{5})/2$ and enter it as a constant $\boxed{\text{K}}$ or into a memory $\boxed{\text{STO}}$ or fill up the stack by repeatedly depressing $\boxed{\text{enter}}$.

(ii) Divide by $\sqrt{5}$. This gives the approximation to u_1.

(iii) Multiply repeatedly by α to generate successively the approximations to u_2, u_3, ...

We turn finally in this discussion of Fibonacci numbers to consider the ratio of successive terms of the sequence:

$$\frac{1}{1}, \frac{2}{1}, \frac{3}{2}, \frac{5}{3}, \frac{8}{5}, \ldots, \frac{u_{n+1}}{u_n}, \ldots$$

We know that

$$u_n = \frac{\alpha^n - \beta^n}{\sqrt{5}} \quad \text{and} \quad u_{n+1} = \frac{\alpha^{n+1} - \beta^{n+1}}{\sqrt{5}}$$

$$\therefore \quad \frac{u_{n+1}}{u_n} = \frac{\alpha^{n+1} - \beta^{n+1}}{\alpha^n - \beta^n}$$

Now we have already shown that $\beta^n \to 0$ as $n \to \infty$ and is always relatively small, so

$$\frac{u_{n+1}}{u_n} \approx \frac{\alpha^{n+1}}{\alpha^n}$$

i.e.

$$\frac{u_{n+1}}{u_n} \approx \alpha \quad \text{and indeed} \quad \frac{u_{n+1}}{u_n} \to \alpha \text{ as } n \to \infty$$

So the ratio of successive terms tends towards $1 \cdot 618 \ldots$ as $n \to \infty$.

Some things to do

Exercise 4.26   Find u_{20} using the Binet formula.

Exercise 4.27   Compute the successive ratios

$$\frac{u_2}{u_1}, \frac{u_3}{u_2}, \frac{u_4}{u_3}, \ldots$$

and show that they do tend to $1 \cdot 618 \ldots$ a number with interesting properties relating to the Golden Section (see F. W. Land, 1975).

Exercise 4.28   Plot the graph of u_n (y-axis) against n (x-axis) for $1 \leqslant n \leqslant 20$, $1 \leqslant u_n \leqslant 7000$. What kind of curve do you get?

GENERATING RECURRING DECIMAL EXPANSIONS

As was mentioned earlier the limited accuracy and display capacity of a pocket calculator mean that it is not possible to find the recurring decimal expansions of many fractions simply by performing the division or even by uncovering hidden digits. A more general algorithm is needed.

It is useful to look at the basis of long division and the example 8/7 will be used.

$$\begin{array}{r} 1{\cdot}14 \\ \hline 7\,)\,\overline{8{\cdot}0000} \end{array}$$

7	Stage 1	7 goes into 8 once, 1 times 7 is 7
10	Stage 2	8 − 7 is 1, bring down the next digit,
7	Stage 3	7 goes into 10 once, 1 times 7 is 7
30	Stage 4	10 − 7 is 3, bring down next digit
28	Stage 5	7 goes into 30 four times, 4 times 7 is 28
20	Stage 6	30 − 28 is 2, bring down next digit

and so on

The important point to note is that at each even numbered stage the appropriate remainder is formed, and is then used for the following division stage. This method is based on the fact that division can be considered as repeated subtraction.

There is no way in which the calculator can be of much use here if long division is performed digit by digit. However there is no reason why only *one digit* at a time should be added to the answer, and the calculator can very much speed up finding decimal expansions in many cases with a method closely paralleling the one above.

First we work through a particular example, 1/17, assuming a calculator with 6-digit display is used, although the method is quite general:

	Operation	*Display*	*Digits in solution*
Stage 1	$1 \div 17$	·058824	·05882

Comment   We ignore the last digit displayed which may be incorrect due to rounding off.

Stage 2	$1 - (17 \times \cdot05882)$	·00006	·05882

Comment   Remainder calculated.

Stage 3	$6 \div 17$	·352941	·0588235294

Comment   Remainder used in division, last digit is ignored. Underlined figures added to solution.

Stage 4	$6 - (17 \times \cdot35294)$	·00002	·0588235294

Comment   Next remainder calculated.

(*continued*)

	Operation	Display	Digits in solution
Stage 5	$2 \div 17$	·117647	·058823529411764

Comment   Next remainder used in division, last digit ignored. More figures added.

Stage 6	$2 - (17 \times \cdot 11764)$	·00012	·058823529411764

Comment   Next remainder calculated.

Stage 7	$12 \div 17$	·705882	·0588235294117 64 70588

Comment   Next remainder used in division, last digit ignored. More figures added.

At this point we see that the digits pattern is recurring. We can learn from number theory that the period (number of digits in the repeating pattern) must be less than the denominator. In this case 16 is the actual period, which is the maximum it could be (denominator 17).

Thus

$$\frac{1}{17} = 0 \cdot \dot{0}58823529411764\dot{7}$$

We can now outline the general procedure to follow when seeking to establish the decimal expansion of any fraction A/B. (see fig. 4.3 p. 69).

Some things to do

Exercise 4.29   Find the recurring decimal expansions of

$$\frac{1}{23}, \quad \frac{1}{26}, \quad \frac{35}{29}, \quad \frac{20}{31}.$$

Exercise 4.30   Investigate the relationship between denominator and period of recurring decimal expansions of fractions of the form $1/n$ where $n = 1, 2, 3, \ldots$.

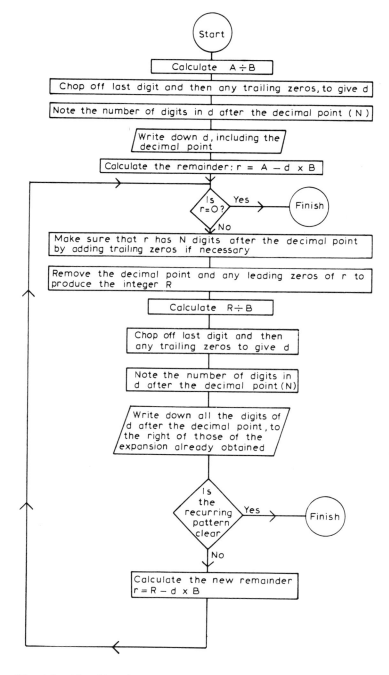

Fig. 4.3   Algorithm for generating recurring decimal expansions.

5. Growth and decay

INTRODUCTION

Although the reader may not feel at all interested in studying the investment of money there is in fact a strong connection with the mathematics it involves and the sciences. A study of interest rates can be a useful introduction to systems which exhibit growth or decay behaviour—common enough phenomena throughout the scientific world.

SIMPLE INTEREST

When a sum of money is invested, for example in a bank or building society, interest is paid to the lender. The *rate* of interest is generally a fixed percentage of the sum invested. If £200 is invested for 1 year at 8% per annum, at the end of the year the interest earned will be £200 × 0·08 = £16. What happens now if the £200 is left invested for a further year? Obviously it will earn another £16—and so on for every year that it remains invested at a rate of 8% p.a. This is an example of a type of investment which earns *Simple Interest*.

An example might be to see how much a principal (that is the sum invested) of $250 invested for 4 years at 7% p.a. simple interest will be worth at the end of the period. Interest earned each year is $250 × 0·07 = $17.50, so for 4 years it will be 4 × $17.50 = $70. Therefore the principal will have grown to $250 + $70 = $320.

The general formula for working out the total interest (I_s) earned by an amount P earning $R\%$ simple interest per period after T periods is

$$I_s = P \times \frac{R}{100} \times T = \frac{PRT}{100}$$

The formula for the total amount (A_s) to which such a principal grows is

$$A_s = P + \frac{PRT}{100} = P\left(1 + \frac{RT}{100}\right)$$

These two formulae

$$\text{Interest: } I_s = \frac{PRT}{100} \qquad \text{Amount: } A_s = P\left(1 + \frac{RT}{100}\right)$$

are very easy to use with a calculator as we shall now see.

Example: A principal of £287·50 is invested for 5 years at $7\frac{5}{8}\%$ simple interest. To how much will it grow?

We must calculate

$$A_s = 287 \cdot 50 \left(1 + \frac{7\frac{5}{8} \times 5}{100}\right)$$

Algebraic key strokes	Display	Comment	Reverse Polish key strokes
5 ÷ 8 +	0·625	$\frac{5}{8}$	5 enter 8 ÷
7 ×	7·625	$7\frac{5}{8}\,(R)$	7 +
5 ÷ 100 +	0·38125	$RT/100$	5 × 100 ÷
1 ×	1·38125	$1 + RT/100$	1 +
287·5 =	397·10937	$P(1 + RT/100)$	287·5 ×

Thus the principal will grow to about £397.11.

Some things to do

Exercise 5.1   How much interest will £1850 earn in 3 years if it earns $6\frac{1}{4}\%$ p.a.?

Exercise 5.2 $327.50 invested for 4 years at $7\frac{7}{8}\%$ p.a. simple interest grows to what amount?

Exercise 5.3   A young couple invest £710.00 for a total period of 5 years hoping that it will grow to £1000.00 for a house deposit. During the first two years the investment earns $7\frac{3}{4}\%$ p.a. simple interest but unfortunately the rate then drops to $7\frac{1}{4}\%$ p.a. Will they reach their target?

COMPOUND INTEREST

Returning to the £200 invested at 8%. After one year £16.00 interest has been earned, and if instead of pocketing this money the lender *invests that as well* he will have £216 now invested, which is a new principal sum. After a further year the £216 will itself earn 8% interest —i.e. $0 \cdot 08 \times £216 = £17.28$, and the sum will have grown to £233.28. This invested for a further year will become £251.94 ... and so on. Such investments where the interest is itself reinvested are said to earn *Compound Interest*.

Is there a quick way to calculate the total amount from year to year?

$$£200 \text{ earns } 0·08 × £200 \text{ to become } £200 + 0·08 × £200$$
$$= £200(1 + 0·08)$$
$$= 1·08 × £200$$

$$£216 \text{ earns } 0·08 × £216 \text{ to become } £216 + 0·08 × £216$$
$$= 1·08 × £216$$
$$= 1·08 × 1·08 × £200$$
$$= (1·08)^2 × £200$$

This is clearly the general pattern, so if the rate is $R\%$, in 1 year P earns $(R/100) × P$ to become

$$P + \frac{R}{100} × P = \left(1 + \frac{R}{100}\right) × P$$

in T years P will grow to

$$\left(1 + \frac{R}{100}\right) × \left(1 + \frac{R}{100}\right) × \ldots \left(1 + \frac{R}{100}\right) × P = \left(1 + \frac{R}{100}\right)^T P$$

∴   An amount P earning $R\%$ p.a. compound interest, whose interest is added annually, will grow in T years to

$$P\left(1 + \frac{R}{100}\right)^T$$

Example:

How much will 233 rupees grow to when earning $6\frac{1}{2}\%$ compound interest for 6 years? How much interest will be earned?

We need to find

$$233 × \left(1 + \frac{6\frac{1}{2}}{100}\right)^6$$

Algebraic key strokes	*Display*	*Comment*	*Reverse Polish key strokes*
6·5 ÷ 100 +	0·065	$R/100$	6·5 enter 100 ÷
1 =	1·065	$1 + R/100$	1 +
y^x 6 =	1·4591423	$(1 + R/100)^T$	6 y^x
× 233 =	339·98016	$P(1 + R/100)^T$	233 ×

Thus the Rs 233 will grow to about Rs 339.98. The interest earned is simply Rs 339.98 − Rs 233.00 = Rs 106.98.

We have been referring to *annual* interest rates and *annual* adding of the interest to the principal and *years* of investment. However the formula developed is applicable to *any period*. Also the interest earned is simply found, by subtracting the original principal invested from the final amount.

The general formula, then, for the amount A_c to which a principal P grows after T periods when earning $R\%$ per period compound interest, where the interest is added ('compounded' or 'compounded up') at the end of each period is

$$A_c = P\left[\left(1 + \frac{R}{100}\right)^T\right]$$

and the interest earned is

$$I_c = P\left[\left(1 + \frac{R}{100}\right)^T - 1\right]$$

$Interest: I_c = P\left[\left(1 + \dfrac{R}{100}\right)^T - 1\right]$	$Amount: A_c = P\left[\left(1 + \dfrac{R}{100}\right)^T\right]$

COMPARISON OF SIMPLE AND COMPOUND INTEREST

If we compare the two methods of investment we can draw up Table 5.1,

TABLE 5.1

	Start	Year 1	Year 2	Year 3	Year 4
Simple	200	$200 + 16$	$200 + 2 \times 16$	$200 + 3 \times 16$	$200 + 4 \times 16$
Compound	200	$200(1 \cdot 08)$	$200(1 \cdot 08)^2$	$200(1 \cdot 08)^3$	$200(1 \cdot 08)^4$

or in terms of amounts to which the money has grown (Table 5.2).

TABLE 5.2

	Start	Year 1	Year 2	Year 3	Year 4
Simple	200	216	232	248	264
Compound	200	216	233·28	251·94	272·10

Plotting these values on a graph we see the nature of the two growth curves (fig. 5.1)

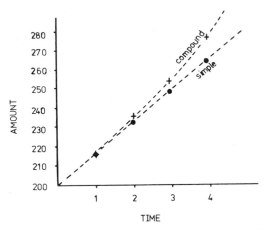

Fig. 5.1   Capital growth under simple and compound interest.

We studied in Chapter 4 arithmetic and geometric progressions (A.P.s and G.P.s). For an A.P. a constant amount is *added* on each time. This is what happens in the Simple Interest case:

Simple Interest   200,   200 + 16,   200 + 2 × 16,   200 + 3 × 16, . . .

A.P. a, $a+d$, $a+2d$, $a+3d$, . . .

clearly $a = 200$ and $d = 16$ in this case.

More generally: $a = P$, the principal invested

$$d = \frac{PR}{100},\text{ the interest earned per period.}$$

For a G.P. a constant *multiplies* each term. This is what happens in Compound Interest:

Compound
Interest   200,   200(1·08),   200(1·08)²   200(1·08)³, . . .

G.P. a, ar, ar^2, ar^3, . . .

clearly $a = 200$, $r = 1·08$ in this case.

More generally: $a = P$, the principal originally invested

$$r = \left(1 + \frac{R}{100}\right),\text{ the growth factor.}$$

Some things to do

Exercise 5.4   To how much will £327.50 grow if invested for 4 years at $7\frac{7}{8}\%$ p.a. compound interest? (Compare with Exercise 5.2.)

Exercise 5.5 Find the interest earned by \$2950 invested for 7 years at $4\frac{1}{2}\%$ p.a. compound interest.

Exercise 5.6 Which earns more—

(a) £100.00 invested for 5 years at 9% p.a.   *or*

(b) £100.00 invested for 9 years at 5% p.a.?

Answer this for both simple and compound interest cases. Firstly try to decide for yourself and then check by calculator. Is the result the same whatever Principal, Time and Rate are taken?

Exercise 5.7 A test tube contains an estimated 1000 bacteria at 12 noon which reproduce by binary fission (i.e. double by splitting) on average every 25 minutes. What will the population be at 5 p.m.?

Exercise 5.8 Repeat Exercise 5.3 assuming that the investment earns compound interest. Will the young couple reach their target in this case?

Exercise 5.9 What annual compound interest rate is equivalent over 5 years to a simple interest rate of $7\frac{1}{2}\%$ p.a.?

BUILDING SOCIETIES

Building Society practice is often not quite as straightforward as the annual compounding of interest. For example, the Leicester Building Society compound the interest twice a year (July 1st, January 1st).

We will find the amount of growth of £100 invested for 1 year at 7% p.a. compound interest, interest compounded every six months. Two friends, Jones and Bright, each decide to invest £100 for a year. Jones invests his money on January 1st 1979 in a Building Society whose compounding dates are January 1st and July 1st.

The formula $A_c = P\left(1 + \dfrac{R}{100}\right)^T$ can be directly used:

$P = 100$

$R = 3\cdot5\%$ for each period of 6 months

$T = 2$ periods of 6 months

$A_c = £100(1 + 0\cdot035)^2 = £107.12$

So one year later Jones gets back £107.12 and the interest rate is actually $7\cdot12\%$.

Bright also invests his money on January 1st 1979 but in a Building Society whose compounding up dates are April 1st and October 1st. For the period January 1st to April 1st (i.e. 3 months or $\frac{1}{4}$ year) the interest rate is $\frac{1}{4} \times 7\% = 1\cdot75\%$ so the £100 will earn £1.75 to be added to the account on April 1st, making £101.75. This is now invested for the 6 months April 1st to October 1st at $3\cdot5\%$ becoming £101.75 $\times$ $1\cdot035 = £105.31$.

Now this £105.31 is invested for 3 months at $1·75\%$ and then on withdrawal the interest due will be added on to give £105.31 × $1·0175 =$ *£107.15* and the interest rate is actually $7·15\%$.

Bright therefore gets back more than Jones much to their puzzlement!

It is seen that for both Jones and Bright the true rate of interest is slightly higher than the advertised ('nominal') rate of 7% p.a. and this is due to the more frequent compounding. We shall return to this later in the chapter where we shall look into what happens if the periodic compounding of the interest is even more frequent.

REGULAR SAVINGS WITH ANNUAL COMPOUNDING

Building Societies try to encourage regular saving by offering a higher rate of interest to customers who deposit a fixed amount each month into an investment account. In effect this is a set of compound interest problems with a new Principal each month.

An example will illustrate the method of calculating the growth of the money.

An investor deposits £20 on the 1st of each month, beginning January 1st 1977, with a Building Society offering $6\frac{1}{2}\%$ p.a. compound interest compounded annually. He continues this investment for 5 years (60 payments). How much should he receive on January 1st 1982?

Take the first year:

£20 will be invested for 12 months, earning £20 × $0·065 = £1·30$

£20 will be invested for 11 months, earning £20 × $11/12$ × $0·065 = £1·19$
etc.

£20 will be invested for 1 month, earning £20 × $1/12$ × $0·065 = £0·11$

After 1 year, then, the total interest will be

£20 × $(1 + 11/12 + 10/12 + \ldots + 1/12)$ × $0·065 = $ £20 × $6·5$ × $0·065 = £8.45$

The total amount at the commencement of the second year will be:

$$12 × £20 + £8.45 = £248.45$$

In the second year this £248·45 will become £248.45 × 1.065 and another £248.45 will be added through the further twelve deposits plus interest. After year 2 we have £248.45 + £248.45(1·065). Similarly in year 3 we have a new set of deposits plus interest together with the year 2 total times 1·065.

$$£248.45 + £248.45(1·065) + £248.45(1·065)^2$$

and after 5 years

$$£248.45(1 + 1·065 + 1·065^2 + 1·065^3 + 1·065^4)$$

This is a G.P. with sum

$$£248.45 \left(\frac{1 \cdot 065^5 - 1}{1 \cdot 065 - 1} \right) = £248.45 \left(\frac{1 \cdot 065^5 - 1}{0 \cdot 065} \right) = £1414.59$$

Generalising the above we have

$$\text{Amount} = M \left(12 + \frac{13}{2} \times \frac{R}{100} \right) \left(\frac{\left[1 + \frac{R}{100} \right]^T - 1}{\frac{R}{100}} \right)$$

where M is the monthly deposit

R is the yearly interest rate (percentage)

T is the number of years for which the account runs.

Something to do

Exercise 5.10 Find the total amount accumulated in an account after 7 years where monthly payments of £23 are made, earning $6\frac{1}{4}\%$ p.a. compounded annually.

REGULAR SAVINGS, THE GENERAL CASE

A modification is needed if the interest is compounded more frequently than annually. Taking the generalized formula:

$$\text{Amount} = M \left(k + \frac{(k+1)}{2} \times \frac{R}{100} \right) \left(\frac{\left[1 + \frac{R}{100} \right]^{12T/k} - 1}{\frac{R}{100}} \right)$$

where M is the monthly deposit

k is the number of months before compounding up occurs

R is the interest rate per k months (percentage)

T is the number of *years* the account runs.

For the frequently used case of six-monthly compounding up $k = 6$ and the example of £20 per month for 5 years, at $6\frac{1}{2}\%$ p.a. gives

$$\text{Amount} = 20(6 + 3 \cdot 5 \times 0 \cdot 0325) \left(\frac{1 \cdot 0325^{10} - 1}{0 \cdot 0325} \right)$$

$$= 122 \cdot 275 \left(\frac{1 \cdot 0325^{10} - 1}{0 \cdot 0325} \right)$$

$$= £1417.99$$

Again we see that the amount is rather larger than for annual compounding up (£1414.59) and so the effective rate of interest is slightly higher than the $6\frac{1}{2}\%$ nominal rate.

Some things to do

Exercise 5.11   (*a*) Check the accuracy of the following table (interest compounded six-monthly) which was in effect in January 1978:

Leicester Building Society

Regular Savings Accounts

This table shows to the nearest £ how your capital will accumulate with regular monthly savings, assuming payments are made punctually each month and the current 6·75% interest rate remains unchanged throughout the periods shown.

The maximum holding in a Regular Savings account is restricted to £3,000.

Monthly Savings	Capital accumulated at the end of:		
	3 years	5 years	7 years
£5	£199	£356	£534
£10	£399	£712	£1068
£15	£598	£1068	£1602
£20	£798	£1423	£2136
£25	£997	£1779	£2671

Fig. 5.2   Typical building society chart with six-monthly compounding.

(b) Can you explain any discrepancy between your results and those of the Leicester Building Society? (fig. 5.2).

Exercise 5.12   Some Building Societies used to compound up monthly. What would £20 per month for 5 years at $6\frac{1}{2}\%$ p.a. grow to in that case?

Exercise 5.13   The Leeds Permanent Building Society in November 1976 offered 9·55% interest rate on Subscription Shares, with annual compounding, and provided the following table. Check its accuracy. (fig. 5.3).

TABLE SHOWING THE ACCUMULATION OF SAVINGS, SUBJECT TO THE CONTINUANCE OF INTEREST AT THE RATE OF 9·55%

Calendar Monthly Payment	BALANCE IN ACCOUNT (including interest accrued)		
	In three years	In five years	In ten years
£2	£ 83·17	£ 152·69	£ 393·61
£4	£ 166·33	£ 305·38	£ 787·22
£8	£ 332·66	£ 610·76	£1574·44
£16	£ 665·33	£1221·52	£3148·88
£20	£ 831·66	£1526·90	£3936·10
£32 (Max.)	£1330·66	£2443·04	£6297·76

Fig. 5.3   Building society interest chart with yearly compounding.

MORTGAGE CALCULATIONS

It is unlikely that the reader, if he is a student at school or college, has much interest in or experience of mortgages. However anyone who sets out to buy a house is almost certain to require a loan for its purchase (called a mortgage) with repayments, usually monthly, over a number of years (10, 15, 20, 25, . . .). Very probably the reader's relatives or friends have mortgages on their houses and it is doubtful whether any of them knows how the relevant calculations are made.

Consider the case of Paul and Rachel, who, on January 1st, 1977 take out a £5000 mortgage, repayable over 20 years, with stated interest rate $10\frac{1}{2}\%$ p.a. How much per month will they have to pay? Building Societies, who are by far the largest lenders of money for mortgages, operate a system whereby at the end of each year the interest due on the Loan (or Principal) for the year is added to the Principal itself, and the repayments made during the year are deducted and the balance is the new Principal for the next year.

For the first year the interest due on L will be $L \times R/100$ so L will grow to $L(1 + R/100) = L(1 + r)$ where $r = R/100$ for simplicity. The repayments will be $M + M + \ldots + M = 12M$. The balance will be $L(1 + r) - 12M$.

This then, is the new amount effectively borrowed for the second year:

$$[L(1 + r) - 12M] \text{ will grow to } [L(1 + r) - 12M] \times (1 + r)$$

and repayments will be $12M$ again so the balance after 2 years is

$$L(1 + r)^2 - 12M(1 + r) - 12M.$$

Similarly after T complete years we have

$$\text{Balance} = L(1 + r)^T - 12M(1 + r)^{T-1} - 12M(1 + r)^{T-2} - \ldots - 12M(1 + r) - 12M.$$

However after T years all the debt should be exactly wiped out and the balance reduced to zero. We therefore have the equation

$$L(1 + r)^T - 12M(1 + r)^{T-1} - \ldots - 12M = 0$$

$$\therefore \quad L(1 + r)^T = 12M[1 + (1 + r) + (1 + r)^2 + \ldots + (1 + r)^{T-1}]$$

The R.H.S. is a G.P. which can be summed by the formula to give

$$L(1 + r)^T = 12M \left[\frac{(1 + r)^T - 1}{r} \right]$$

So the formula for monthly repayments is

$$M = \frac{rL(1 + r)^T}{12[(1 + r)^T - 1]}$$

We can now calculate how much Paul and Rachel must pay per month.

M is the required monthly repayment (to be found)

$R = 10.5\%$ so $r = 0.105$

$L = £5000$

$T = 20$ years

$$\therefore \quad M = \frac{0.105 \times 5000 \times (1.105)^{20}}{12[(1.105)^{20} - 1]}$$

$$= £50.622194.$$

So we might take the monthly repayments as £50.62 or £50.63. Another rather shattering calculation can now be made. How much will they actually repay over the 20 years for the loan of £5000?

$$20 \times 12 \times £50.62 = £12148.80$$

In the early years of paying off a mortgage most of each repayment goes towards repaying the interest on the loan rather than repaying the capital (initial amount borrowed) so the amount owed hardly decreases at all at first as Table 5.3 shows:

TABLE 5.3   *Mortgage repayments*

Year	Amount owed at beginning of year	Interest paid during year	Capital repaid during year
	£	£	£
1	5000.00	525.00	82.44
2	4917.56	516·34	91.10
3	4826.46	506.78	100·66
4	4725.80	496.21	111.23
5	4614.57	484.53	122.91
6	4491.66	471.62	135.82
7	4355.84	457.36	150.08
8	4205.76	441.60	165.84
9	4039.92	424.19	183.25
10	3856.67	404.95	202.49
11	3654.18	383.69	223.75
12	3430.43	360.20	247.24
13	3183.19	334.23	273.21
14	2909.98	305.55	301.89
15	2608.09	273.85	333.59
16	2274.50	238.82	368.62
17	1905.88	200.12	407.32
18	1498.56	157.35	450.09
19	1048.47	110.09	497.35
20	551.12	57.87	549.57
21	1.55		

Note that the amount owed after the 20 years are up is £1.55 rather than £0.00. This is because £50.62 is slightly less than the required theoretical monthly payment, and also rounding off each year's figures may cause a slight deviation from the result of using the formula which assumes that no rounding off takes place.

The report after their first year which Paul and Rachel might receive would appear as:

Amount of Loan at 1/1/77	..	..	£5000.00
Interest due at 31/12/77 ..	..	..	£525.00
	Total	..	£5525.00
Repayments made up to 31/12/77	..	..	−£607.44
	Balance	..	£4917.56

The astute reader may have noticed that the monthly repayments do not earn any interest although they are in effect invested with the Building Society until the 'reckoning up' at the end of each year. This really means that the true rate of interest is somewhat higher than the published figure.

Some things to do

Exercise 5.14   What should the monthly repayments be on a mortgage of £7950.00 over 25 years if the rate is $9\frac{7}{8}\%$ p.a.?

Exercise 5.15   A mortgage is arranged for £4600 with a lending rate of $9\frac{1}{4}\%$ p.a. If the monthly repayments are £76.80 how long will it be before the debt is cleared?

> *Hint:* This can be done by computing a mortgage table like Table 5.3 for Paul and Rachel's mortgage or by rearranging the equation relating M, r, L, T to make T the subject of the formula. The first is perhaps easier but the second is of more general use of course.

Exercise 5.16   Recalculate the first column of the table for Paul's and Rachel's mortgage (Table 5.3) taking £50.63 as the monthly repayment. Compare with the figures in Table 5.3.

Exercise 5.17   Investigate the rate of interest being charged in 1976 by the Cavendish Finance Company for mortgage loans (see fig. 5.4).

'INSTANT INTEREST'

In our discussions of interest we have only considered cases where interest is compounded annually or six-monthly. What if the interest is calculated more frequently? The pocket calculator can be put to good use in investigating this.

For example:

£100 is invested for 1 year at 6% compound interest. If the interest is added after 12 months it will just be $£100 \times 0.06 = £6.00$. However, with 6-monthly compounding:

The 6% p.a. effectively becomes 3% per 6 months:

$$A = 100.00(1 + 6/[2 \times 100])^2 = £106.09$$

and an extra 9 pence is thereby earned!

What about if compounding up is done 4 times a year?

$$A = 100.00(1 + 6/[4 \times 100])^4 = £106.14$$

and if done 12 times a year?

$$A = 100.00(1 + 6/[12 \times 100])^{12} = 106.17$$

It seems clear that we will get more interest the more rapidly is the compounding. Is there a limit or can the £100.00 earn an indefinitely

Fig. 5.4   Advertisement for loans.

large amount of interest? What about if the compounding up is done daily?

$$A = 100 \cdot 00(1 + 6/[365 \times 100])^{365} = £106.18$$

And hourly?

$$A = 100 \cdot 00(1 + 6/[24 \times 365 \times 100])^{24 \times 365} = £106.18$$

And every minute?

$$A = 100 \cdot 00(1 + 6/[60 \times 24 \times 365 \times 100])^{60 \times 24 \times 365} = £106.18$$

And there does seem to be a limit to what can be earned—equivalent to about 6·18% p.a.

Something to do

Exercise 5.18 A Manhattan bank as a publicity stunt offers $7\frac{1}{2}\%$ p.a. nominal interest rate with 'instant compounding'. A second more sober bank offers $7\frac{3}{4}\%$ p.a. interest compounded annually. Which is the better deal? If the interest rates of both banks drop by 1% is there any change in the situation?

THE EXPONENTIAL FUNCTION EMERGES

The previous investigation into 'instant interest' led us to find out what happens to $(1+0\cdot06/n)^n$ as n gets larger and larger (i.e. as 'n tends to infinity' which is symbolized by $n\rightarrow\infty$). Although not very meaningful in terms of money growth the case where the $0\cdot06$ is replaced by 1 (i.e. 100% interest rate) is applicable to other growth situations (e.g. bacteria cultures) and also leads to some important mathematical ideas.

What happens, then, to $(1+1/n)^n$ as $n\rightarrow\infty$? This is easily investigated with the aid of a calculator:

Taking $n=100$, then $n=1000$ etc., for example:

Algebraic key strokes	*Display*	*Reverse Polish key strokes*
100 $\boxed{1/x}$ $\boxed{+}$ 1 $\boxed{=}$	1·01	100 $\boxed{1/x}$ 1 $\boxed{+}$
$\boxed{y^x}$ 100 $\boxed{=}$	2·7048138	100 $\boxed{y^x}$
1000 $\boxed{1/x}$ $\boxed{+}$ 1 $\boxed{=}$	1·001	1000 $\boxed{1/x}$ 1 $\boxed{+}$
$\boxed{y^x}$ 1000 $\boxed{=}$	2·7169239	1000 $\boxed{y^x}$

A table of values (Table 5.4) can be calculated (the reader may like to check this):

TABLE 5.4

n	$1/n$	$(1+1/n)^n$
1	1	2
10	0·1	2·5937425
10^2	0·01	2·7048138
10^3	0·001	2·7169239
10^4	0·0001	2·7181459
10^5	0·00001	2·7182682
10^6	0·000001	2·7182805
10^7	0·0000001	2·7182818

If a calculator with more significant figures is used we find

$n=10^8$ gives 2·718281828 and so does $n=10^9$.

We have in fact found in Table 5.4 the limit of $(1+1/n)^n$ as $n\rightarrow\infty$

to 7 d.p. to be 2·7182818. Expressed in mathematical symbols

$$\text{Lim}_{n\to\infty} (1+1/n)^n = 2\cdot7182818 \text{ to 7 d.p.}$$

This limit is the well known number called e, which is the base of 'natural' or 'Napierian' logarithms. Like π (3·14159 . . .) and $\sqrt{2}$ (1·4142 . . .) it is a never-ending decimal which does not have a recurring pattern. Unlike $\sqrt{2}$ however, π and e are not the roots of any algebraic equations and so π and e are called *transcendental* numbers. Functions involving these numbers (e.g. trigonometric, hyperbolic, logarithmic, exponential) are called transcendental functions.

The case considered, $(1+1/n)^n$, is rather special, (corresponding to a 100% interest rate!) and a more general result comes from $(1+x/n)^n$ which can represent *any* interest rate by suitable choice of x. Is there a similar limit to $(1+x/n)^n$ if $n\to\infty$ for a given x? (The reader could investigate this for various choices of x.)

It turns out that

$$\text{Lim}_{n\to\infty} (1+x/n)^n = e^x \quad \text{where } e = 2\cdot71828 \ldots$$

This function, e^x, is known as the exponential function, and is sometimes expressed as exp (x). All scientific calculators include this function, the key being marked $\boxed{e^x}$.

Returning to our problem of evaluating the 6% 'instant interest' investment of £100 the calculation required is

$$100\left(1+\frac{6}{n\times100}\right)^n = 100\left(1+\frac{0\cdot06}{n}\right)^n$$

where n is the number of times the compounding up is done. If n is allowed to increase without limitation then 'instant interest' is obtained, and the necessary expression is

$$\text{Lim}_{n\to\infty}\left\{100\left(1+\frac{6}{n\times100}\right)^n\right\}$$

$$= 100\,\text{Lim}_{n\to\infty}\left\{\left(1+\frac{0\cdot06}{n}\right)^n\right\}$$

$$= 100 \times e^{0\cdot06} \quad \text{(from the definition of } e^x\text{)}$$

$$= 100 \times 1\cdot0618365 = 106\cdot18365 \ldots \text{ evaluated on a pocket calculator.}$$

Thus £100 will grow to £106.18 to the nearest penny, a result which we had already discovered by taking large values for n when the topic was first looked at. We have incidentally found that there certainly is a limit to the amount to which a sum grows with 'instant interest'. Limits are discussed more fully in Chapter 9.

THE EXPONENTIAL FUNCTION

It was shown earlier that Simple and Compound Interest yields could be graphed as in fig. 5.1 (page 74).

The dots and crosses are the meaningful points and the lines merely join them up to improve the look of the graph, but intermediate positions between dots or crosses may have no obvious meaning.

As the frequency of compounding up is increased the compound curves have more and more dots on them, and the curves get slightly higher, i.e. the increase is greater (fig. 5.5).

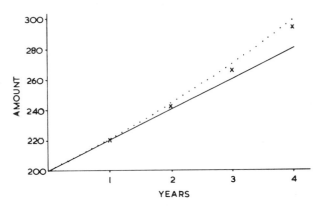

Fig. 5.5   Capital growth under different kinds of interest.

Key: ╱   Simple Interest.
      ×   Compound Interest, annually compounded.
     ....   Compound Interest, compounded 10 times per annum.

If compounding up is made instantaneous we get a *continuous* curve corresponding to e^{kx} for some positive value of k.

The graph is continuous because there are dots for all the instants of time whereas the compound graphs are really discrete (not continuous) consisting of a set of separated dots.

Mathematicians generalize the definition of the exponential function, e^{kx}, allowing negative x values too, and a series expansion has been shown to be equivalent to it, thus:

$$e^{kx} = 1 + \frac{kx}{1} + \frac{(kx)^2}{1 \times 2} + \frac{(kx)^3}{1 \times 2 \times 3} + \frac{(kx)^4}{1 \times 2 \times 3 \times 4} + \cdots$$

for all x, and the corresponding graph, for $k > 0$, is shown in fig. 5.6.

Many natural processes exhibit the exponential curve associated with compound interest and geometric progressions. In favourable conditions a population of N bacteria might grow by binary fission, to

2N after 30 minutes. The size, at 30 minute intervals, would be approximately N, 2N, 4N, 8N, 16N, ... 2$^r N$, ... which is clearly a G.P.

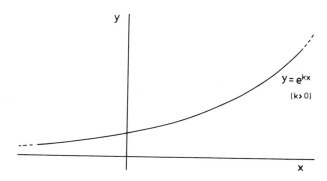

Fig. 5.6   Exponential growth.

Some things to do

Exercise 5.19   Compute $e^{0\cdot1}$ using the expansion

$$e^x = 1 + x + \frac{x^2}{1\times2} + \frac{x^3}{1\times2\times3} + \ldots$$

to machine accuracy and compare with the function value (1·1051709181 to 10 d.p.).

Exercise 5.20   It can be shown that the curves $y = e^{kx}$ and $y = a^x$ are exactly the same if $k = \ln(a)$—see 'Solutions and Notes' for an explanation. For example, taking $a = 2$, $\ln(2) = 0\cdot69314718$ to 8 d.p. so $y = e^{0\cdot69314718x}$ is the same as $y = 2^x$.

Check this by computing values to complete the table below, using the $\boxed{e^x}$ and $\boxed{y^x}$ keys.

x	-4	$-1\cdot5$	0	$2\cdot5$	10	$12\cdot3$
$y = 2^x$						
$y = e^{0\cdot69314718x}$						

DECAY CURVES

So far we have concentrated on examples which tend to exhibit an exponential growth pattern, increasing with time.

There are other situations in which the graph falls instead of rising, and the basic mathematical model for this is still $y = e^{kx}$, but this time

$k < 0$. It is the exponential function again but because k is negative e^{kx} gets *smaller* as x gets *larger*.

Some things to do

Use a calculator to complete the following tables and hence sketch the curves and verify that they exhibit patterns of growth and decay.

Exercise 5.21 $y = 2e^{0.2x}$

x	0	1	2	3	4	5	6	7	8	9	10
$y = 2e^{0.2x}$											

Exercise 5.22 $y = 100e^{-0.01x}$

x	0	50	100	150	200	250	300	350	400	450	500
$y = 100e^{-0.01x}$											

DECAY MODELS

The Decay curve occurs in various physical situations, for example the forced cooling of a hot body, first order chemical reactions, radioactive decay of isotopes, leakage of capacitors. Such physical cases have mathematical models (or formulae) involving e^{kt} where t is time and k is a negative constant.

Example

Newton's Law of Cooling states that the rate of loss of heat from a hot body in a strong steady draught of air is proportional to the temperature difference between the air and the cooling body.

This leads to the differential equation:

$$\frac{d\theta}{dt} = -k(\theta - \theta_a)$$

where θ_a is the air temperature

θ is the body temperature

t is the time

k is some positive constant

The solution to this differential equation is $\theta = \theta_a + (\theta_0 - \theta_a)e^{-kt}$, where θ_0 is the initial temperature of the hot body.

A car radiator becomes over-heated to 100°C and the driver stops. The radiator is cooled by the battery-driven fan blowing air at 20°C. Plot the graph to show the fall in temperature with time, taking $k = 0.05$.

To do this we first draw up a table of values and use a calculator for the necessary computations:

$$\theta_a = 20, \quad \theta_0 = 100, \quad k = 0.05$$

so $\theta = 20 + 80e^{-0.05t}$ is the equation connecting temperature θ and time t.

We shall take $t = 5, 10, 15, \ldots$ minutes.

Algebraic key strokes	*Display*	*Comment*	*Reverse Polish key strokes*
·05 $\boxed{+/-}$ $\boxed{\times}$			·05 $\boxed{+/-}$ $\boxed{\text{enter}}$
5 $\boxed{=}$ $\boxed{e^x}$	·77880078	$e^{-0.25}$	5 $\boxed{\times}$ $\boxed{e^x}$
$\boxed{\times}$ 80 $\boxed{+}$ 20			80 $\boxed{\times}$ 20
$\boxed{=}$	82·304063	θ at 5 min.	$\boxed{+}$
·05 $\boxed{+/-}$ $\boxed{\times}$			·05 $\boxed{+/-}$ $\boxed{\text{enter}}$
10 $\boxed{=}$ $\boxed{e^x}$	·60653066	$e^{-0.5}$	10 $\boxed{\times}$ $\boxed{e^x}$
$\boxed{\times}$ 80 $\boxed{+}$ 20			80 $\boxed{\times}$ 20
$\boxed{=}$	68·522453	θ at 10 min.	$\boxed{+}$
and so on	....	....	and so on

A table of values is thus produced (Table 5.5) from which a graph (fig. 5.7) can be sketched.

TABLE 5.5

Time t	0	5	10	15	20	25	30	35	40	45	50	55	60
Temp. T	100	82	69	58	49	43	38	34	31	28	27	25	24

The following examples illustrate the wide variety of situations in which the exponential function occurs, and indicate how useful it is to have the $\boxed{e^x}$ key on a calculator.

Some things to do

Exercise 5.23　The main limitation of the exponential growth curve as representing population growth is that it does not tail off as other factors (e.g. overcrowding, food shortage, predation) become important with increasing numbers.

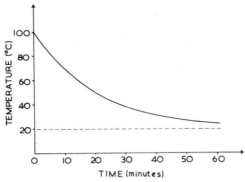

Fig. 5.7   Cooling of a car radiator (see Table 5.5).

A mathematical model which *does* allow for this is the *logistic curve* with equation:

$$P=\frac{P_0be^{bt}}{b-dP_0+dP_0e^{bt}}$$

where P_0 represents the starting size of the population and b and d are birth rate and death rate constants which determine how the curve of growth is shaped for a particular situation.

One hundred rabbits are released on a small island and their breeding pattern leads to a population growth represented by the logistic curve where $b=0.3$ and $d=0.001$ and time t is measured in years. Plot the growth of the population at 2-yearly intervals for 20 years. What does the maximum size of the population appear to be?

Exercise 5.24   Compare the logistic curve $y=1/(1+e^{-x})$ with the Gompertz curve $y=e^{-e^{-x}}$ for $-3\leqslant x\leqslant3$ by computing sufficient values to sketch the two curves on the same graph.

Exercise 5.25   The radioactive decay of a nuclide is described by the equation

$$N=N_0e^{-\lambda t}$$

where N_0 is the initial number of nuclei, N is the number at time t and λ is a decay constant depending on the particular nuclide and on the units in which t is measured.

For ^{228}actinium $\lambda\approx0.1131$ where t is measured in hours. Use the calculator to provide values to plot the graph of the decay of ^{228}Ac (to form ^{228}Th) for 12-hourly intervals assuming that initially $N_0=10^6$.

From the graph estimate the *half-life* $T_{\frac{1}{2}}$ of ^{228}Ac—i.e. the time taken for half of the original number of nuclei present to decay.

Compare with the theoretical half-life found by the direct calculation $T_{\frac{1}{2}}=\ln(2)/\lambda$.

Exercise 5.26   The atmospheric pressure P at any place is given by

$$P=P_0e^{-kH/T}.$$

where P_0 is the air pressure at sea level, Pa

$\quad$ k is a constant, 0.0343 km^{-1}

$\quad$ T is the air temperature, K

$\quad$ H is the height above sea level, m.

Find the atmospheric pressure at the top of Mount Everest (8848 m), when the temperature is $-15°C$ and the air pressure at sea level is 1.01×10^5 Pa.

Exercise 5.27 Uranium 234 (^{234}U) disintegrates into thorium 230 (^{230}Th) which in turn decays to radium 226 (^{226}Ra) which decays relatively rapidly through a chain of reactions to lead (^{206}Pb). Schematically we have:

$$^{234}U \rightarrow {}^{230}Th \rightarrow {}^{226}Ra \xrightarrow[\text{through several steps}]{\text{rapidly}} {}^{206}Pb$$

The half lives are 235000 years for ^{234}U, and 80000 years for ^{230}Th.

If we wish to predict the build-up of the 'daughter' product, thorium, we need to take account of its generation from the 'mother', uranium, and its natural disintegration. The appropriate formula for N_T, the mass of thorium at time t, is:

$$N_T = \frac{\lambda_U}{\lambda_T - \lambda_U} N_0 (e^{-\lambda_U t} - e^{-\lambda_T t})$$

where N_0 is the mass of uranium to start with,

$$\lambda_U = \frac{\ln(2)}{235000}, \quad \lambda_T = \frac{\ln(2)}{80000}$$

By taking the case $N_0 = 193.75$g the constant term $\lambda_U/(\lambda_T - \lambda_U) N_0$ becomes 100 so the formula then simplifies to

$$N_T = 100(e^{-\lambda_U t} - e^{-\lambda_T t})$$

(a) Find λ_U and λ_T and then complete the table below and hence plot the graph of the abundance of the thorium against time.

Time (1000's years)	0	25	50	75	100	150	200	300
N_T g	0							

Time (1000's years)	400	500	600	700	800	900	1000
N_T g							

(b) The uranium itself decays according to the formula $N_U = N_0 e^{-\lambda_U t}$ (where $N_0 = 193 \cdot 75$g).

On the same graph as for (a) plot the abundance of the uranium against time.

Exercise 5.28   Suppose we have to evaluate $1/e^{-230}$. If we try to do so by first calculating 230 $\boxed{+/-}$ $\boxed{e^x}$ we get *underflow*—the number e^{-230} is too small to be represented on the calculator and an error condition arises or a zero is produced. In either case we cannot form the required reciprocal. How can the answer be obtained?

6. The scientific functions in action

We have already met the exponential curve $y=e^x$ in Chapter 5 (fig. 5.6). If the inverse of this function is wanted, it can be obtained graphically by reflecting the e^x curve in the line $y=x$ (fig. 6.1). This curve is the natural logarithm curve

$$y=\log_e x=\ln x$$

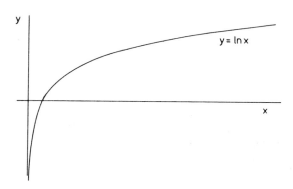

Fig. 6.1   Natural logarithm, ln x.

which only exists for positive x and keeps rising as x increases but does so more and more slowly. [We shall use ln x for the natural logarithm and log x for the 'common' logarithm, $\log_{10} x$.]

The e^x function can be thought of as the 'antilogarithm' for natural logarithms, which work to base e ($=2\cdot71828\ldots.$).

The logarithm of a number is the *power* (or index) needed to produce that number by raising the *base* to the power. E.g.

ln $20=2\cdot9957323$ to 8 s.f. because $e^{2\cdot9957323}=20$

log $20=1\cdot3010300$ to 8 s.f. because $10^{1\cdot3010300}=20.$

Something to do

Exercise 6.1 The following properties are true for logarithms to *any* base. Check them for ln and log, choosing your own values for A, B, n.

(a) $\log(A) + \log(B) = \log(A \times B)$ (b) $\log(A) - \log(B) = \log(A/B)$

(c) $\log(A^n) = n\log(A)$

APPLICATIONS OF NATURAL LOGARITHMS

When a gas is compressed isothermally (i.e. keeping the temperature constant) from a pressure p_1 and volume V_1 to a smaller volume V_2 the external work done on the gas is

$$W = p_1 V_1 \ln\left(\frac{V_1}{V_2}\right)$$

What is the external work done in compressing a certain mass of gas isothermally from 20×10^{-4} m^3 at $16 \cdot 2 \times 10^4$ Pa to 7×10^{-4} m^3?

$$W = 16 \cdot 2 \times 10^4 \times 20 \times 10^{-4} \times \ln\left(\frac{20}{7}\right)$$

$$= 310 \cdot 14237 \text{ J}.$$

So the work done is about 340 J.

Radiocarbon dating is a method often used to date ancient objects containing carbon taken from the atmosphere. The assumption is that the amount of ^{14}C in the object will have begun to decay radioactively when the intake of carbon ceased—usually on the death of living organisms in the object—from an initial level of 15·3 counts per minute per gram of carbon. Knowing the current radio-active level and the law of radio-active decay and the half life, $T_{\frac{1}{2}}$, of ^{14}C, it is possible to approximately date an object (up to about 40000 years of age perhaps). The formula is:

$$\text{Age} = \frac{T_{\frac{1}{2}}}{\ln 2} \times \ln\left(\frac{N_c}{N}\right)$$

where $N_c = 15 \cdot 3$ c.p.m.

$N =$ current c.p.m.

$T_{\frac{1}{2}} =$ half life of ^{14}C, 5650 years approximately.

For example the bones of an extinct reptile are found to give a count of 3·7 c.p.m. When was the animal alive?

$$\text{Age} = \frac{5650}{\ln(2)} \times \ln\left(\frac{15 \cdot 3}{3 \cdot 7}\right)$$

$$= 11570 \text{ years}$$

So the reptile bones are about $11\frac{1}{2}$ thousand years old.

Some things to do

Exercise 6.2   The entropy change ΔS joules for 1 mole of a gas is given by

$$\Delta S = c_p \ln (T_2/T_1) + R \ln (p_1/p_2)$$

where T_1 and T_2 are initial and final temperatures (K)

    p_1 and p_2 are initial and final pressures (pascals)

    c_p is the heat capacity of the gas at constant pressure (J K^{-1} mol^{-1})

    R is the gas constant, 8·314 J K^{-1} mol^{-1}.

Find the entropy change involved in converting 1 mole of chlorine from $2·3 \times 10^5$ pascals at 298 K to $1·1 \times 10^5$ pascals at 320 K, given that $c_p \approx 33·7$ J K^{-1} mol^{-1}.

Exercise 6.3   An important numerical constant (rather like π and e) is Euler's γ, which is useful in some arrangement and probability problems and crops up occasionally in advanced mathematics.

It can be defined as

$$\gamma = 1 + \frac{1}{2} + \frac{1}{3} + \frac{1}{4} + \ldots + \frac{1}{n} - \ln n \quad \text{where } n \to \infty.$$

This is particularly interesting as it is the difference of two infinite quantities. By taking some values of n calculate some approximations to γ, whose value is $0·5772\ldots$.

Exercise 6·4   Looked at another way, the result of Exercise 6·3 can provide us with an approximate formula for the harmonic series

$$\sum_{1}^{n} \frac{1}{r} = 1 + \frac{1}{2} + \frac{1}{3} + \ldots + \frac{1}{n}$$
$$\approx 0·5772 + \ln (n)$$

Use this method to estimate the following:

$$\sum_{1}^{50} \frac{1}{r}, \quad \sum_{1}^{250} \frac{1}{r}, \quad \sum_{1}^{10^6} \frac{1}{r}$$

Exercise 6.5   Estimate the age of a scroll which produces 11·9 counts per minute per gram of carbon.

EXPONENTIAL AND LOGARITHMIC FUNCTIONS IN GENERAL

$y = e^x$ is an important special case of the more general exponential function $y = a^x$ where a is any positive number, whose graph is as in fig. 6.2.

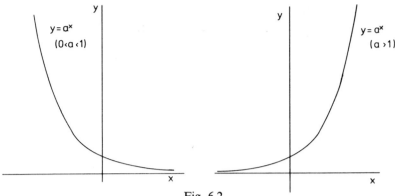

Fig. 6.2

As was indicated in Exercise 5.20 of Chapter 5 $y=a^x$ is precisely the same as $y=e^{kx}$ where $k=\ln a$ so all exponential functions $y=a^x$ can be thought of as $y=e^{kx}$ for suitable choice of k.

If a problem arises involving a^x then the $\boxed{y^x}$ key on the calculator is useful rather than converting to e^{kx} form and using the $\boxed{e^x}$ key.

For example, when the pressure p and volume V of a fixed mass of ideal gas change under adiabatic conditions (i.e. no heat is gained or lost) the relationship between p and V is given by

$$pV^\gamma = \text{constant}$$

where γ is a constant (the ratio of the principal specific heat capacities), which depends on the atomicity of the gas.

For air $\gamma \approx 1\cdot41$. Given 20 litres of air at $1\cdot4\times10^5$ Pa, what will be the pressure if the volume it occupies is reduced adiabatically to 13 litres?

Here we use

$$p_1 V_1{}^\gamma = p_2 V_2{}^\gamma$$

$p_1 = 1\cdot4\times10^5$, $V_1 = 20$, p_2 is to be found, $V_2 = 13$, $\gamma = 1\cdot41$

$$\therefore \quad p_2 = p_1 \frac{V_1{}^\gamma}{V_2{}^\gamma} = p_1 \left(\frac{V_1}{V_2}\right)^\gamma = 1\cdot4\times10^5 \times \left(\frac{20}{13}\right)^{1\cdot41}$$

Algebraic key strokes	*Display*	*Comment*	*Reverse Polish key strokes*
20	20	V_1	20 $\boxed{\text{enter}}$
$\boxed{\div}$ 13 $\boxed{=}$	1·5384615	V_1/V_2	13 $\boxed{\div}$
$\boxed{y^x}$ 1·41 $\boxed{\times}$	1·8356597	$(V_1/V_2)^{1\cdot41}$	1·41 $\boxed{y^x}$
140000 $\boxed{=}$	256992·35	p_2	140000 $\boxed{\times}$

so $p_2 \approx 2\cdot57 \times 10^5$ Pa.

LOGARITHMS TO DIFFERENT BASES

There is a relationship between logarithms in different bases, *a* and *b*.

$$\log_a N = \frac{\log_b N}{\log_b a}$$

For example, to find $\log_3 201 \cdot 5$ we might calculate

$$\frac{\log_e 201 \cdot 5}{\log_e 3} = \frac{5 \cdot 3057894}{1 \cdot 0986123}$$

$$= 4 \cdot 8295376$$

making use of the $\boxed{\ln}$ function on the calculator.

Some things to do

Exercise 6.6   A monatomic gas (for which $\gamma \approx 1 \cdot 66$) is contained in a vessel at a pressure of $2 \cdot 15 \times 10^5$ Pa in a volume of $16 \cdot 2$ litres. Find the pressure when the gas is compressed adiabatically to a volume of $12 \cdot 1$ litres.

Exercise 6.7   The formula, due to Shannon, for the channel capacity, *C*, of a telecommunications system is

$$C = 2W \log_2 (1 + S/N)$$

where *W* is the bandwidth and S/N is the signal-to-noise ratio.
   Find *C* if $W = 3600$ and $S/N = 18$.

Exercise 6.8   The normal human body temperature is 37°C. On death the temperature gradually falls to that of the surroundings. If it is assumed that the temperature, θ, falls according to Newton's law of cooling then if the temperature of the surroundings, θ_s, is measured and the body temperature at two separate times *t* minutes apart is taken (θ_1, θ_2) the time of death can be estimated, using the formula

$$\text{time of death} \approx \frac{\ln \left\{ \dfrac{\theta_1 - \theta_s}{37 - \theta_s} \right\}}{\ln \left\{ \dfrac{\theta_2 - \theta_s}{\theta_1 - \theta_s} \right\}} \, t \quad \text{minutes before the first reading } (\theta_1).$$

   An executive is found murdered in his air-conditioned office (constant temperature 20°C). The body's temperature is 29·5°C when it is discovered at 3 p.m. 110 minutes later the body's temperature has fallen to 27°C. Estimate the time of death.

Exercise 6.9   Newton's law of cooling really only applies if the body is in a strong constant draught. A more appropriate formula assumes that the rate of cooling is proportional to $(\theta - \theta_s)^{1 \cdot 3}$ instead of $(\theta - \theta_s)^{1 \cdot 0}$ which Newton's law uses. The formula is this case is:

$$\text{time of death} \approx \frac{(37-\theta_s)^{-\cdot3}-(\theta_1-\theta_s)^{-\cdot3}}{(\theta_1-\theta_s)^{-\cdot3}-(\theta_2-\theta_s)^{-\cdot3}}\, t \quad \begin{array}{l}\text{minutes before the}\\ \text{first reading } (\theta_1).\end{array}$$

Repeat Exercise 6.8 using this second formula. Does it make enough difference to help the Police with their enquiries?

COMMON LOGARITHMS

Just as $\ln x$ and e^x are inverses of each other, so too are the functions $\log_{10} x$ and 10^x.

Common logarithm tables simply provide some particular values of the curve $y=\log_{10} x$ (fig. 6.3).

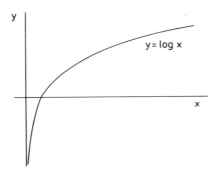

Fig. 6.3　Common logarithm, log x.

The antilogarithm tables correspond to $y=10^x$.

The connection between the natural and common logarithms of a number, N is:

$$\boxed{\log_{10} N = \log_{10} e \times \log_e N}$$

Now $\log_{10} e = 0\cdot43429448$ to 8 d.p. so $\log_{10} N \approx 0\cdot43429448 \ln N$.

Example:

$$\log_{10} 23 \approx 0\cdot43429448 \times \ln 23$$
$$\approx 0\cdot43429448 \times 3\cdot1354942$$
$$\approx 1\cdot3617278$$

which is the value obtained using the $\boxed{\log}$ function directly. This simple connection between logarithms of the different bases means that not having base 10 logarithms on a calculator is no loss. In any case common logarithms are little used other than in pencil-and-paper calculations and the pocket calculator replaces logs for that purpose anyway.

It is not even necessary to remember the 0·43429448 factor because an alternative form of the relationship (already given in the general case on page 97) is:

$$\log_{10} N = \frac{\log_e N}{\log_e 10}$$

so the whole computation can be performed on the calculator using the $\boxed{\ln}$ function. For example:

Find $\log_{10} 279\cdot31$.

Algebraic key strokes	Display	Comment	Reverse Polish key strokes
279·31 $\boxed{\ln}$	5·6323223	ln (N)	279·31 $\boxed{\ln}$
$\boxed{\div}$ 10 $\boxed{\ln}$	2·3025851	ln (10)	10 $\boxed{\ln}$
$\boxed{=}$	2·4460865	ln (N)/ln (10)	$\boxed{\div}$

APPLICATIONS OF COMMON LOGARITHMS

The acidity or alkalinity of a solution is dependent on the concentration of hydrogen ions in the solution. It is expressed in terms of pH value:

$$pH = -\log [\text{concentration of } H^+ \text{ in mol } l^{-1}]$$

For example a solution with concentration 0·00037 mol l^{-1} has pH

$$-\log 0\cdot00037 = 3\cdot432 \text{ to 3 d.p.}$$

A pH of less than 7 is acidic, 7 is neutral and greater than 7 is alkaline.

Some things to do

Exercise 6.10 A formula connecting the mass in kilograms (W), the height in metres (H) and the body surface area in square metres (A) for an adult human has been given by Dreyer as

$$\log A = 0\cdot725 \log H + 0\cdot425 \log W - 0\cdot6937$$

What is the approximate surface area of a person of mass 80 kg and of height 1·75 m?

Exercise 6.11 Find the pH of the following aqueous solutions:

(a) 0·047 mol l^{-1} [H^+]   (b) 0·0000000031 mol l^{-1} [H^+].

Exercise 6.12 The numbers of registered private cars in Great Britain for the last thirty years are shown in the table below:

Year	1946	1947	1948	1949	1950	1951	1952	1953
Millions of cars	1·77	1·94	1·96	2·13	2·26	2·38	2·51	2·76

Year	1954	1955	1956	1957	1958	1959	1960	1961
Millions of cars	3·10	3·53	3·89	4·19	4·55	4·97	5·53	5·98

Year	1962	1963	1964	1965	1966	1967	1968	1969
Millions of cars	6·56	7·38	8·25	8·92	9·51	10·30	10·82	11·29

Year	1970	1971	1972	1973	1974	1975	1976
Millions of cars	11·52	12·06	12·72	13·50	13·64	13·75	14·03

[*Source:* Annual Abstract of Statistics, Central Statistical Office, H.M.S.O., London.]

Is this an exponential growth pattern? It is hard to be sure by plotting the graph as to whether the curve is of the form $y = a^t$. The easiest graph to recognize is the straight line. By plotting the logarithm of the car numbers ($\log y$) against time (t) a straight line will result if the relationship is exponential. Use this method to test the data.

THE DIFFERENCE BETWEEN USING LOG TABLES AND USING A CALCULATOR

It is the usual practice when using log tables to express the log of a number smaller than 1 in 'mixed form'.

Example:

$$\log_{10} 0{\cdot}2345 = \bar{1}{\cdot}3701$$

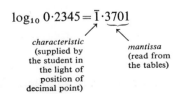

characteristic
(supplied by
the student in
the light of
position of
decimal point)

mantissa
(read from
the tables)

The 'bar one' ($\bar{1}$) really means -1 but the mantissa (·3701) is *positive* so the logarithm is actually

$$-1 + \cdot3701 = -\cdot6299 \text{ to 4 d.p.}$$

What happens when we use a calculator? We get directly $-\cdot62985715$, i.e. the actual log (not in mixed form).

This is not any problem because for use in formulae this is how we do want the log to appear. The mixed notation is solely a device to simplify (!) using log tables in pencil-and-paper calculations.

Should you wish to find the mixed form of a log the process is quite simple:

Example:
$$-1\cdot1919241 = -2 + \cdot8080759$$
$$= \bar{2}\cdot8080759$$

THE TRIGONOMETRIC FUNCTIONS

The trigonometric functions sin, cos, tan; and their reciprocals cosec, sec, cot are commonly met in mathematics. Scientific calculators do not carry the reciprocal functions as they are so simply found from the first three.

For example, to find cosec 27° which is 1/sin 27° the key-stroke sequence would be:

27 $\boxed{\text{sin}}$ $\boxed{1/x}$ giving 2·2026893.

However

27 $\boxed{\text{sin}}$ $\boxed{1/x}$ can also give 1·0456139!

Why is that? This is a timely warning to check whether the calculator is working in *degrees* or *radians*, depending on the setting of a small switch on the calculator or on the pressing of a special function key.

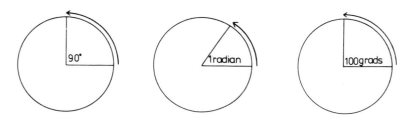

Fig. 6.4 Three different types of unit for measuring angles.

A circle is divided in 360 degrees (360°) so a right angle is 90° (fig. 6.4). A bigger unit is the radian (rad). There are 2π (i.e. $2 \times 3 \cdot 14159 \ldots$) radians in a circle, so 1 radian is $360°/2\pi \approx 57 \cdot 3°$.

A third measure of angle sometimes met is the grad. There are 400 grads in a circle—so a grad is slightly smaller than a degree. This system of angle units means that a right angle is 100 grads.

A second kind of error easily committed is in calculations such as $\cos (2A)$ where $A = 30°$, say. The algebraic key-stroke sequence 2 $\boxed{\times}$ 30 $\boxed{\cos}$ $\boxed{=}$ yields $1 \cdot 7320508$ on many machines, which is not a very likely value for *any* cosine! The error here is in forgetting that $\boxed{\cos}$ acts solely on what is in the display (i.e. 30) and the multiplication 2 $\boxed{\times}$ 30 is not yet performed. The correct key sequence is 2 $\boxed{\times}$ 30 $\boxed{=}$ $\boxed{\cos}$ yielding $0 \cdot 5$. Users with Reverse Polish calculators are less likely to make this particular kind of error. Some algebraic logic machines will produce the result $0 \cdot 8660254$ because they lose the 2 $\boxed{\times}$ part when the $\boxed{\cos}$ has been pressed, which is very unfortunate. It is to be hoped that such machines are not produced in the future.

Example:

When a projectile is fired over level ground with a velocity u at an angle α (fig. 6.5) ignoring the effects of air resistance the range, R, is given by

$$R = \frac{u^2 \sin 2\alpha}{g}.$$

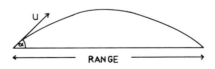

Fig. 6.5

The magnitude of the actual velocity, V, at any time, t, during the period of flight is

$$V = \sqrt{(u^2 \cos^2 \alpha + (u \sin \alpha - gt)^2)}$$

For a shell fired with initial velocity of $109 \cdot 6 \text{ ms}^{-1}$ at an angle of $36 \cdot 4°$, find (i) the range (ii) the actual speed of the shell after $3 \cdot 5$ seconds.

(i) We use $R = u^2 \sin 2\alpha/g$ where $u = 109 \cdot 6$, $g = 9 \cdot 81$, $\alpha = 36 \cdot 4°$.

Algebraic key strokes	Display	Comment	Reverse Polish key strokes
Select degree mode			Select degree mode
36·4	36·4	α	36·4 [enter]
[×] 2 [=]	72·8	2α	2 [×]
[sin]	·95527836	$\sin 2\alpha$	[sin]
[×] 109·6 [x²]	12012·16	u^2	109·6 [x²]
[÷] 9·81 [=]	1169·7203	R	[×] 9·81 [÷]

Thus the range of the shell is about 1170 metres.

(ii) We use $V = \sqrt{(u^2 \cos^2 \alpha + (u \sin \alpha - gt)^2)}$ where $u = 109\cdot6$, $g = 9\cdot81$, $\alpha = 36\cdot4°$, $t = 3\cdot5$.

Algebraic key strokes	Display	Comment
Select degree mode		
9·81 [×] 3·5 [=] [STO]	34·335	gt
109·6 [×] 36·4 [sin]	·59341889	$\sin \alpha$
[−] [RCL] [=] [x²] [STO]	942·71781	$(u \sin \alpha - gt)^2$
109·6 [x²]	12012·16	u^2
[×] 36·4 [cos] [x²]	·64785403	$\cos^2 \alpha$
[+] [RCL] [=] [√x]	93·406873	V

Note: the key-stroke sequence using brackets is much simpler.

Reverse Polish key strokes	Display	Comment
Select degree mode		
109·6 [enter] [enter]	109·6	u (stored in stack)
36·4 [cos] [×] [x²]	7782·1262	$(u \cos \alpha)^2$
[x⇌y]	109·6	u (recovered from stack)
36·4 [sin] [×]	65·03871	$u \sin \alpha$
9·81 [enter]	9·81	g
3·5 [×] [−] [x²]	942·71781	$(u \sin \alpha - gt)^2$
[+]	8724·844	$(u \cos \alpha)^2 + (u \sin \alpha - gt)^2$
[√x]	93·406873	V

Thus the shell is travelling at about 93·4 ms^{-1} after $3\frac{1}{2}$ seconds.

Some things to do

Exercise 6.13 (fig. 6.6)   (*a*) A person of normal sight can just read print which has apparent height of about $\frac{1}{12}°$—i.e. the image makes an angle, θ, of $\frac{1}{12}°$ to the eye. The general trigonometric formula connecting d, h, θ is $h = d \tan \theta$.

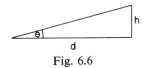

Fig. 6.6

Show that the height, h, of print which can be read by the normally sighted person at a distance $d = 25$ m is about 3·6 cm.

(*b*) For small angles $\tan \theta$ can be replaced by θ (provided θ is in radians), giving in this example the simpler formula $h = d \times \theta$.

Check that this applies to the above problem.

(*c*) How far away can lettering of height 22 cm be read?

Exercise 6.14   The height of a projectile in flight is

$$h = x \tan \alpha - \frac{gx^2}{2u^2 \cos^2 \alpha}$$

where α is the angle of projection,

u is the speed of projection,

g is the acceleration due to gravity,

x is the horizontal distance from the projection point.

A shell is fired with a velocity 200 ms^{-1} at an angle of 17·5° aimed to hit an ammunition dump 2340 m away.

(*a*) Check as to whether the angle is correct to give the required range ($R = u^2 \sin 2\alpha/g$).

(*b*) Will the shell clear an intervening block of flats of height 65 m situated 2100 m from the firing position?

Exercise 6.15   A student uses a pocket calculator to evaluate $\sin \theta$ for a particular value of $\theta°$. A friend tries the same calculation, keying in the same number, but by mistake assumes θ to be in radians. However the two get the same answer! What is the angle θ?

Exercise 6.16   S. J. Briggs (1971) has analysed the mechanics of toy motor cars ('hot wheels') which race along ramps, loop the loop, and so on.

One piece of apparatus consists of a long ramp down which a car races and jumps over a gap to land on a second inclined ramp (fig. 6.7).

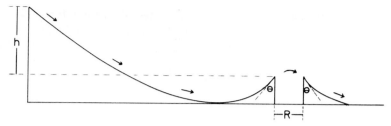

Fig. 6.7 Path of toy car, which runs down a ramp and jumps a gap.

The height, h, at which the car must be released above the point where it leaves the first ramp, is given by

$$h = \frac{R}{2e \sin 2\theta}$$

where R is the horizontal distance to be jumped and e is the efficiency of the car ($e = 1$ for no friction).

Briggs found experimentally that $e \approx 0.8$.

Find the height needed to jump a gap of 30 cm where the ramps are angled at 17° ($\theta = 73°$).

Exercise 6.17 John Bernoulli set a problem in 1696 as follows. 'Find the shape of the frictionless path down which a body should fall under gravity from point O to point P such that the time of descent is a minimum.' No continental mathematician could solve it in the six months set but it took Isaac Newton no more than 12 hours when he learned of it!

The answer to this, the 'brachistochrone problem', is an inverted *cycloid* (see fig. 6.8) with equation:

$$x = r(\theta - \sin \theta), \quad y = r(1 - \cos \theta)$$

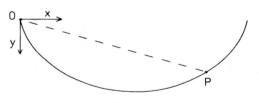

Fig. 6.8

It is the path traced out by a point on the circumference of a wheel radius r as it travels along a straight path turning through an angle θ radians about its own axis.

It is interesting to compare the times of descent from O to P travelling along the curve and along the straight line joining O and P.

For the cycloid the time is $T_c = \theta\sqrt{(r/g)}$ where r, θ are the parameters of the point P.

For the straight line path the time is

$$T_1 = \sqrt{\left(\frac{2r(1-\cos\theta)}{g} \cdot \frac{1+k^2}{k^2} \right)}$$

where

$$k = \frac{1-\cos\theta}{\theta-\sin\theta}$$

which is the downward gradient of OP.

Complete the following table and check that the cycloid path is always less. What is the physical explanation? How might friction affect the comparisons?

Note that r and g only occur as $\sqrt{(r/g)}$ in both expressions and so do not affect the comparison. Choosing $r=9.8$ m, $g=9.8$ ms^{-2} simplifies the working:

$$T_c = \theta; \quad T_1 = \sqrt{\left(2(1-\cos\theta)\frac{1+k^2}{k^2} \right)}$$

$\theta/$rad	$\pi/4$	$2\pi/4$	$3\pi/4$	$4\pi/4$	$5\pi/4$	$6\pi/4$	$7\pi/4$
$T_c/$s	0·785	1·571	2·356	3·142	3·927	4·712	5·498
$T_1/$s	0·792						

(A version of this problem is discussed by J. C. Siddons (1976).)

INVERSE TRIGONOMETRIC FUNCTIONS

Given a value for x we can easily compute $y = \sin x$, e.g. if $x = 0.563$ radians, $y = \sin 0.563 = 0.53372557$.

However, what if we are told what y is and need to find x? e.g. $0.621 = \sin x$, what is x?

This is an inverse problem: we require $x = \sin^{-1} 0.621$ or $x = $ arcsin 0.621.

Most scientific calculators have inverse trigonometric functions available—often a separate $\boxed{\text{arc}}$ key is pressed before the relevant trigonometric key.

Example:

Key strokes	*Display*	*Comment*
(a) Select degree mode		Degrees chosen
·621	·621	
arc sin	38·389197	answer in degrees
(b) Select radian mode		Radians chosen
·5	·5	
arc cos	1·0471976	answer in radians

It should be noted that arcsin x and arccos x are only defined for $-1 \leqslant x \leqslant 1$ because sines and cosines are restricted to this range of values but arctan (x) has no such restriction. Also there are really infinitely many solutions to these inverse function problems because the trigonometric functions are periodic:

For example   sin $30° = 0·5$
                sin $150° = 0·5$
                sin $390° = 0·5$
                and so on.

The calculator will always give the *principal value*, which is the smallest angle (Table 6.1).

TABLE 6.1

Function	Domain for x	Range for answer in radians	Range for answer in degrees
$y = \arcsin x$	$-1 \leqslant x \leqslant 1$	$-\dfrac{\pi}{2} \leqslant y \leqslant \dfrac{\pi}{2}$	$-90 \leqslant y \leqslant 90$
$y = \arccos x$	$-1 \leqslant x \leqslant 1$	$0 \leqslant y \leqslant \pi$	$0 \leqslant y \leqslant 180$
$y = \arctan x$	no restriction	$-\dfrac{\pi}{2} \leqslant y \leqslant \dfrac{\pi}{2}$	$-90 \leqslant y \leqslant 90$

It is very rarely that the other inverse trigonometric functions are needed. (They can be derived from the basic ones.)

The inverse trigonometric functions can be useful to do quick conversions from degrees to radians (or grads) and vice versa. Supposing we wish to know what $75°$ is in radians:

Key strokes	*Display*	*Comment*
Select degree mode		
75	75	angle in degrees
sin	·96592583	sin $75°$
Select radian mode		
arc sin	1·3089969	angle in radians

Some things to do

Exercise 6.18   The shortest distance between two points on the Earth's surface (assuming it to be a sphere) is that along a Great Circle.

If A has Latitude θ_A and Longitude ϕ_A and B has Latitude θ_B and Longitude ϕ_B the formula is

distance $= 60$ arccos $[\sin \theta_A \sin \theta_B$
$$+ \cos \theta_A \cos \theta_B \cos (\phi_B - \phi_A)]$$ nautical miles.

Use this formula to determine how far apart are

   (*a*) Edinburgh (55° 57′N, 3° 13′W) and New Delhi (28° 37′N, 77° 13′E).

   (*b*) Washington D.C. (38° 55′N, 77° 00′W) and Adelaide (34° 56′S, 138° 36′E).

[*Note:* North and West are positive, South and East are negative, for the angles of Latitude and Longitude.]

Exercise 6.19   When a car turns a corner the two front wheels must turn through different angles if the steering is to be true. This is achieved approximately using special steering gear (Ackermann, for example). The formula relating width between wheel centres (track) T and distance between axles (wheelbase) B and the two angles α, β is

$$\frac{T}{B} = \cot \alpha - \cot \beta$$

The radius of the turning circle is $R = B/\sin \alpha$ (see fig. 6.9).

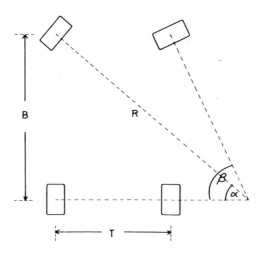

Fig. 6.9   Position of wheels as a vehicle turns a corner.

The BLMC Rover 3500 has a track of 1500 mm and wheelbase of 2815 mm and a minimum turning radius of 5220 mm.

Use $\alpha = \arcsin (B/R)$ and then

$\beta = \arctan (1/(\cot \alpha - T/B))$ to find the angles through which the wheels must turn when on full lock.

Exercise 6.20   J. C. Siddons (1975) discusses how the refractive index of a substance which produces a bow when light is shone on an array of small spheres (e.g. rainbow) can be calculated directly from a measurement of the full angle, 2θ, subtended at the eye by the bow. The formulae are:

$$i = 2 \arctan [\sqrt[3]{(\tan (\theta/4)}]$$
$$r = \theta/4 + i/2$$

Refractive index $= \sin i / \sin r$.

For a rainbow the angle subtended at the eye, 2θ, is 84°. Estimate the refractive index of water.

COORDINATE SYSTEMS: RECTANGULAR CARTESIAN AND POLAR

The normal 2-dimensional system used is the familiar rectangular cartesian coordinate system.

Another useful system is the polar coordinate system where instead of two distances (x and y) being given to fix a point's position, one distance (r) and one angle (θ) are specified (fig. 6.10).

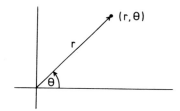

Fig. 6.10

Conversion from polar to rectangular cartesian:

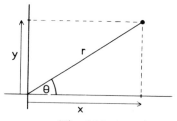

Fig. 6.11

This is quite straightforward. By trigonometry we have (fig. 6.11):

$$\text{(a) } x = r \cos \theta \qquad \text{(b) } y = r \sin \theta$$

Example:

Express (2·5, 139°) in Cartesian form:

Algebraic key strokes	Display	Comment	Reverse Polish key strokes
Select degree mode			Select degree mode
(a) 139 $\boxed{\cos}$ $\boxed{\times}$			139 $\boxed{\text{enter}}$ $\boxed{\cos}$
2·5 $\boxed{=}$	$-1\cdot8867739$	x	2·5 $\boxed{\times}$
(b) 139 $\boxed{\sin}$ $\boxed{\times}$			$\boxed{\text{R}\downarrow}$ $\boxed{\sin}$
2·5 $\boxed{=}$	$1\cdot6401475$	y	2·5 $\boxed{\times}$

Conversion from rectangular cartesian to polar:

This is not quite so straightforward as it requires the $\boxed{\tan^{-1}}$ or $\boxed{\text{arc}}$ $\boxed{\tan}$ function and the procedure needed depends on whether the x value is positive or negative.

By Pythagoras' theorem:

(a) $r = \sqrt{(x^2 + y^2)}$

By trigonometry:

(b) $\tan \theta = y/x$ so

$\qquad \theta = \arctan (y/x)$ *if x is positive*

$\qquad \theta = 180° + \arctan (y/x)$ *if x is negative.*

(The principal value is always given for arctan (y/x), which lies in the range $-90°$ to $+90°$).

Example: See page 111.

Some calculators have a function key to perform these conversions automatically.

Some things to do

Exercise 6.21   The exact formula for the length of belt needed to go round two pulleys of radii R_1, R_2 ($R_1 \geqslant R_2$) whose centres are d apart is:

$$L = 2\pi R_1 - 2(R_1 - R_2) \arccos \left(\frac{R_1 - R_2}{d} \right) + 2\sqrt{(d^2 - (R_1 - R_2)^2)}$$

For various values of R_1, R_2 compare the approximate formula used in Exercise 4.12 with the exact formula. (Can you derive the two formulae yourself?)

Example:

Express $(-1\cdot88, 1\cdot64)$ in polar form:

	Algebraic key strokes	Display	Comment	Reverse Polish key strokes	
(a)	Select degree mode	—	—	Select degree mode	(a)
(b)	1·88 [+/−] [+] 1·64 [x²] [=]	6·224	$x^2 + y^2$	1·88 [+/−] [x²] 1·64 [x²] [+]	(b)
(c)	[√x]	2·4947945	r	[√x]	(c)
(d)	1·64 [÷] 1·88 [+/−] [=]	−·87234042	y/x	1·64 [enter] [+/−] 1·88 [÷]	(d)
(e)	[arc] [tan]	−41·099506	arctan (y/x)	[arc] [tan]	(e)
(f)	[+] 180 [=]	138·90049	θ	180 [+]	(f)

Note 1: The [+/−] is really unnecessary in line (b).

Note 2: Memories or the stack can be used to avoid the need to enter the x and y values twice. (It is an interesting exercise to devise a Reverse Polish scheme using only the stack!)

Exercise 6.22   Convert from polar to rectangular cartesian:

(a) (72·3, 63°)

(b) (4·26, 212° 11′).

Exercise 6.23   Convert from rectangular cartesian to polar:

(a) (−2·7, −7·3)

(b) (6·3, −0·4).

Exercise 6.24   The path of a flying insect moving towards a point source of light has been observed to approximate to the equiangular spiral $r=a^\theta$ where (r, θ) is the polar coordinate system, θ being measured in radians.

Many organisms grow in the form of the equiangular spiral—the Nautilus shell, the arrangement of florets in the heads of flowers (e.g. daisy, sunflower), the horns of many animals, being some examples.

For the case $a=0·8$ complete the table of values, remembering to convert the angle θ to radians:

θ	30°	60°	90°	120°	150°	180°
$r=a^\theta$	0·89	0·79				

θ	210°	240°	270°	300°	330°	360°
$r=a^\theta$						

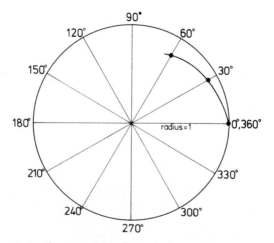

Fig. 6.12   The start of an equiangular spiral, a common curve found in Nature.

and plot the points (r, θ) on a polar graph or convert to cartesian first if you prefer (fig. 6.12). Continue the curve by performing the necessary calculations and directly marking the points on the graph.

F. W. Land (1975) and D'Arcy Thompson (1961) discuss this curve and its biological connections.

VECTOR ADDITION

The speed of an aircraft and the direction of flight together constitute a *vector* quantity, its velocity, (having both magnitude and direction). Other common examples of vector quantities are acceleration, force and displacement (change of position relative to some origin).

Given two vectors (r_1, θ_1) with components (x_1, y_1) and (r_2, θ_2) with components (x_2, y_2) their sum is (R, θ) where

$$R = \sqrt{((x_1 + x_2)^2 + (y_1 + y_2)^2)}$$

$$\theta = \arctan\left(\frac{y_1 + y_2}{x_1 + x_2}\right) \text{ [plus } 180° \text{ if } x_1 + x_2 < 0]$$

$$x_1 = r_1 \cos \theta_1$$
$$x_2 = r_2 \cos \theta_2$$
$$y_1 = r_1 \sin \theta_1$$
$$y_2 = r_2 \sin \theta_2$$

Example

A destroyer sails from its base port a distance of 290 km on a bearing 054° and then changes direction and sails 173 km on a bearing of 156°. What is its final position relative to the base port? (fig. 6.13).

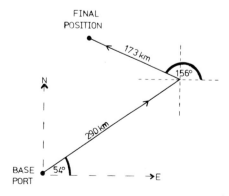

Fig. 6.13   Typical vector addition problem.

Here

$$r_1 = 290, \theta_1 = 54°$$
$$r_2 = 173, \theta_2 = 156°$$

The computation involved is simplified if the polar↔rectangular cartesian conversion facility is available as this will give x_1, y_1, x_2, y_2 straightaway and after some simple calculations will give the final R, θ values. The key-stroke sequences given on page 115 do not use that facility, but assume two memories. Many variations are possible of course, depending on both the machine and the user's preferences.

Some things to do

Exercise 6.25   An aircraft flies at an airspeed of 395 knots on a bearing 312°, in a wind of 83 knots blowing in a direction 163°. Find the true (ground) speed of the aircraft.

Exercise 6.26   Find the resultant of the three concurrent forces

$$u = 14N \text{ acting at } 60°$$
$$v = 12N \text{ acting at } 174°$$
$$w = 13N \text{ acting at } 279°$$

HYPERBOLIC FUNCTIONS

The functions sinh, cosh are used in scientific and engineering applications (e.g. potentials in power lines, shapes of suspended wires) and are of interest to mathematicians. They first arose in studying the properties of the hyperbola, as the name suggests. The two basic definitions are sinh $x = (e^x - e^{-x})/2$, cosh $x = (e^x + e^{-x})/2$ and their ratio is tanh $x = $ sinh $x/$cosh x (fig. 6.14).

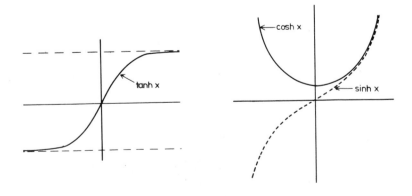

Fig. 6.14   Hyperbolic curves.

Algebraic key strokes	Display	Comment	Reverse Polish key strokes
Select degree mode	—	—	Select degree mode
290 × 54 cos = STO1	170·45772	x_1	290 enter 54 cos ×
173 × 156 cos =	−158·04336	x_2	173 enter 156 cos ×
+ RCL1 = STO1	12·414359	$x_1 + x_2$	+ STO1
290 × 54 sin = STO2	234·61492	y_1	290 enter 54 sin ×
173 × 156 sin =	70·365439	y_2	173 enter 156 sin ×
+ RCL2 = STO2	304·98036	$y_1 + y_2$	+ STO2
x² + RCL1 x² =	93167·14	$(y_1 + y_2)^2 + (x_1 + x_2)^2$	x² RCL1 x² +
√x	305·23292	R	√x
RCL2 ÷ RCL1 =	24·566743	$(y_1 + y_2)/(x_1 + x_2)$	RCL2 RCL1 ÷
arc tan	87·669037	θ	arc tan

As $x_1 + x_2 > 0$ there is no need to add 180° to θ in this example.
The destroyer is *305·2* km from the base on a bearing *087·67°*.

Thus the values are readily calculated using the $\boxed{e^x}$ key if the functions themselves are not provided.

Often a single $\boxed{\text{hyp}}$ key is used in conjunction with the ordinary trigonometric keys to give the basic hyperbolic functions, e.g. $2\cdot79$ $\boxed{\text{hyp}}$ $\boxed{\cos}$ gives $\cosh 2\cdot79 = 8\cdot1712205$ to 8 s.f.

Example

Suppose we know that for some x, $\sinh x = 11013\cdot23287$ and that $\cosh x = 11013\cdot23292$. How can we find e^{-x} (assuming we do not have inverse hyperbolic functions)?

In fact $e^{-x} = \cosh x - \sinh x = 11013\cdot23292 - 11013\cdot23287$.

This gives us $4\cdot53 \times 10^{-5}$ using a 10-digit calculator. Instead of the 10 s.f. in the original data we are now down to 3 s.f. in the answer (only the first two of which are correct).

However

$$e^{-x} = \frac{1}{\cosh x + \sinh x} = \frac{1}{11013\cdot23292 + 11013\cdot232}$$

This gives us $4\cdot539992976 \times 10^{-5}$, a 10 s.f. answer! But is it accurate? Yes—the value of x is in fact $10\cdot0$ and so we can easily check by calculating e^{-10} which gives the same answer. (You may like to check these calculations for yourself.)

This provides us with another very good example of how, by reformulating a problem into a suitable form, much greater accuracy can result by avoiding loss of significant figures by cancellation.

Some things to do

Exercise 6.27   The shape in which a perfectly flexible string will hang when suspended at its two ends is called a catenary. Hanging chains, telegraph wires etc. more-or-less hang in this same curve.

The equation is $y = c \cosh (x/c)$ where c is a constant (called a parameter) which determines the amount of sag in the curve.

Plot the graph of $y = 4 \cosh (x/4)$ for $x = -8$ to 8 in unit steps and check that the curve produced does resemble a hanging chain.

Exercise 6.28   The length, s, along a symmetrical catenary from the lowest point to some arbitrary point where $x = d$ is $s = c \sinh (d/c)$.

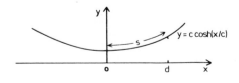

Fig. 6.15

It is required to support a length of telegraph wire between two identical pylons standing on level ground 50 m apart. If the parameter $c = 100$ m how long must the wire be?

INVERSE HYPERBOLIC FUNCTIONS

Given x, we can readily compute $y = \sinh x$. E.g. for $x = 2 \cdot 31$, $y = \sinh 2 \cdot 31 = 4 \cdot 9875817$.

However if we know y can we find the x such that $\sinh x = y$? E.g. for $y = 4 \cdot 567$ what is x? We need to solve $4 \cdot 567 = \sinh x$. This is an inverse problem: we want to find $x = \sinh^{-1} 4 \cdot 567$, or $x = \operatorname{arcsinh} 4 \cdot 567$.

Some calculators have this facility. Usually the $\boxed{\text{arc}}$ key is pressed before pressing the function key $\boxed{\text{sinh}}$ or the pair $\boxed{\text{hyp}}\ \boxed{\text{sin}}$:

Key strokes	Display	Comment
4·567	4·567	x
$\boxed{\text{arc}}\ \boxed{\text{hyp}}\ \boxed{\text{sin}}$	2·2237799	$\sinh^{-1} x$ or $\operatorname{arcsinh} x$

Some things to do

Exercise 6.29   It is stated in calculus books, in connection with integrating $1/\sqrt{(x^2 + a^2)}$ that the result is $\sinh^{-1}(x/a)$ and that this is the same as

$$\ln \left\{ \frac{x + \sqrt{(x^2 + a^2)}}{a} \right\}$$

Check the truth of this for the cases $x = 2$, $a = 1$ and $x = 2 \cdot 3$, $a = 0 \cdot 7$.

Exercise 6.30   The network attenuation, A(dB), of the passive band-pass filter shown in fig. 6.16 can be calculated from

$$A = 40 \log [e^{\operatorname{arcsinh}\sqrt{K}}] \quad \text{for } K > 0$$
$$A = 0 \quad \text{for } 0 \geqslant K \geqslant -1$$
$$A = 40 \log [e^{\operatorname{arccosh}\sqrt{-K}}] \quad \text{for } -1 > K$$

where

$$K = (\omega^2 C_1 L_1 - 1)(1 - \omega^2 C_2 L_2)/(4\omega^2 C_1 L_2)$$

$\omega = 2\pi f$ is the 'angular frequency'

C is the capacitance

L is the inductance.

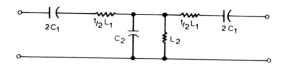

Fig. 6.16

For the cases
$C_1 = 4 \times 10^{-6}F$; $C_2 = 1\cdot3 \times 10^{-5}F$; $L_1 = 3 \times 10^{-2}H$; $L_2 = 4 \times 10^{-3}H$
with (i) $f = 200$ Hz; (ii) $f = 400$ Hz; (iii) $f = 600$ Hz, find the attenuation, in dB.

FOURIER SERIES

It has been shown that almost any function, whether continuous or broken, can be expressed as an infinite sum of sine and cosine terms for any finite interval $[a, b]$. That is, for example the function shown in fig. 6.17 can be written as

$$y = a_0 + a_1 \cos x + a_2 \cos 2x + a_3 \cos 3x + \ldots$$
$$+ b_1 \sin x + b_2 \sin 2x + b_3 \sin 3x + \ldots$$

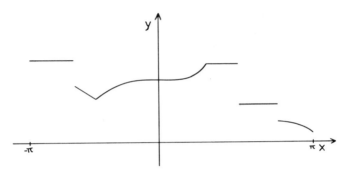

Fig. 6.17   Discontinuous function.

Fourier Series analysis seeks to find the a and b values to fit the particular function.

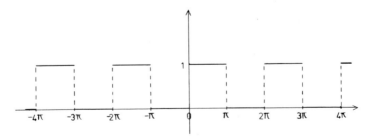

Fig. 6.18   Square-wave function.

The square wave (fig. 6.18) is important in electronics and has as its Fourier Series

$$f(x) = \frac{1}{2} + \frac{2}{\pi}\left[\sin x + \frac{\sin 3x}{3} + \frac{\sin 5x}{5} + \frac{\sin 7x}{7} + \ldots\right]$$

Taking a few terms of this expansion for various values of x will indicate how close to the required value (1 or 0) the series is. The calculator makes such checks manageable whereas using tables is extremely unsatisfactory.

Taking terms up to $\dfrac{2}{\pi}\dfrac{\sin 19x}{19}$

$f(1) \approx \quad 0\cdot992$, should be 1 for the infinite series
$f(2) \approx \quad 1\cdot012$, should be 1 for the infinite series
$f(3) \approx \quad 1\cdot085$, should be 1 for the infinite series
$f(4) \approx -0\cdot003$, should be 0 for the infinite series

Clearly the function's strange behaviour is being reasonably well represented by the truncated series

$$\frac{1}{2}+\frac{2}{\pi}\sin x+\ldots+\frac{2}{\pi}\frac{\sin 19x}{19}.$$

Something to do

Exercise 6.31   The function

$$f(x)=\pi-x \quad \text{for } 0<x<\pi$$
$$=\pi+x \quad \text{for } -\pi<x<0$$

is as shown in fig.6.19.

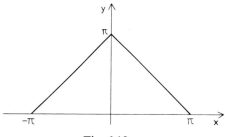

Fig. 6.19

Its Fourier Series is

$$f(x)=\frac{\pi}{2}+\frac{4}{\pi}\left[\cos x+\frac{\cos 3x}{3^2}+\frac{\cos 5x}{5^2}+\frac{\cos 7x}{7^2}+\ldots\right]$$

By taking terms up to $\cos 19x$ in the expansion estimate $f(1)$, which should be $\pi-1$. Also choose one other value of x in the range $-\pi<x<\pi$ and see if $f(x)$ is reasonably well represented by this truncated Fourier Series.

7. Statistics

ARITHMETIC MEAN

The ordinary average of a set of data (the arithmetic mean) is the simplest of the various averages which can be calculated. It can of course be found by adding up all the data values and dividing by how many there are. Obviously a calculator can be useful for this.

Those with the $\boxed{M+}$ or $\boxed{\Sigma}$ key may want to use it for adding up all the terms but this has no advantage in key-stroke economy.

For example, find the (arithmetic) mean of 6, 2, 7, 9, 8.

Using an Algebraic machine we might proceed as follows:

6 $\boxed{STO}$ 2 $\boxed{M+}$ 7 $\boxed{M+}$ 9 $\boxed{M+}$ 8 $\boxed{M+}$ $\boxed{RCL}$ $\boxed{\div}$ 5 $\boxed{=}$

(14 key strokes) whereas more directly we could proceed:

6 $\boxed{+}$ 2 $\boxed{+}$ 7 $\boxed{+}$ 9 $\boxed{+}$ 8 $\boxed{\div}$ 5 $\boxed{=}$

(12 key strokes, but those using 'sum of products' logic machines require an extra $\boxed{=}$ before $\boxed{\div}$).

Why did we use $\boxed{STO}$ first time and not $\boxed{M+}$? ($\boxed{STO}$ clears the memory first but $\boxed{M+}$ does not.)

The Reverse Polish key strokes are

6 $\boxed{enter}$ 2 $\boxed{+}$ 7 $\boxed{+}$ 9 $\boxed{+}$ 8 $\boxed{+}$ 5 $\boxed{\div}$ (12 key strokes)

The reason why the symbol Σ (Greek capital letter sigma) is used on some calculators for the $\boxed{M+}$ key is because

$$\sum_{i=1}^{n} x_i$$

means 'Sum up all the x_i's for i going from 1 to n'—in other words *sum up* $x_1 + x_2 + \ldots + x_n$ so $\boxed{\Sigma}$ is the *summation* key.

The arithmetic mean of n numbers $x_1, x_2 \ldots x_n$ can be mathematically expressed as

$$\text{mean} = \frac{1}{n} \sum_{i=1}^{n} x_i.$$

Frequency tables

Often data are presented not item by item but in a frequency table:

Blackbird egg clutch sizes in a sample of 45 nests

nos. of eggs.	1	2	3	4	5	6	(value)
nos. of nests	2	8	21	13	0	1	(frequency)

The best way to calculate the mean clutch size is to evaluate

$$[2 \times 1 + 8 \times 2 + 21 \times 3 + 13 \times 4 + 0 \times 5 + 1 \times 6] \div 45$$

How you proceed will depend on your particular machine's facilities. We have discussed this before but it is perhaps worth reviewing:

(i) Using a memory:

Algebraic key strokes	Display	Comment
2 $\boxed{\text{STO}}$	2	2×1
8 $\boxed{\times}$ 2 $\boxed{+}$ $\boxed{\text{RCL}}$ $\boxed{=}$ $\boxed{\text{STO}}$	18	8×2 added
21 $\boxed{\times}$ 3 $\boxed{+}$ $\boxed{\text{RCL}}$ $\boxed{=}$ $\boxed{\text{STO}}$	81	21×3 added
13 $\boxed{\times}$ 4 $\boxed{+}$ $\boxed{\text{RCL}}$ $\boxed{+}$	133	13×4 added
6 $\boxed{\div}$ 45 $\boxed{=}$	3·0888889	arithmetic mean calculated

(ii) Using parentheses the procedure is to key stroke directly:

2 $\boxed{+}$ $\boxed{[(}$ 8 $\boxed{\times}$ 2 $\boxed{)]}$ $\boxed{+}$ $\boxed{[(}$ 21 $\boxed{\times}$ 3 $\boxed{)]}$
$\boxed{+}$ $\boxed{[(}$ 13 $\boxed{\times}$ 4 $\boxed{)]}$ $\boxed{+}$ 6 $\boxed{\div}$ 45 $\boxed{=}$

(iii) If $\boxed{\text{M}+}$ or $\boxed{\Sigma}$ is available the key stroking is very simple, providing you remember to clear the memory first (or use $\boxed{\text{STO}}$ initially). Where a machine has several memories it is important to know which memory $\boxed{\Sigma}$ operates on and so address it correctly.

Algebraic:

2 $\boxed{\text{M}+}$ 8 $\boxed{\times}$ 2 $\boxed{=}$ $\boxed{\text{M}+}$ 21 $\boxed{\times}$ 3 $\boxed{=}$ $\boxed{\text{M}+}$
13 $\boxed{\times}$ 4 $\boxed{=}$ $\boxed{\text{M}+}$ 6 $\boxed{\text{M}+}$ $\boxed{\text{RCL}}$ $\boxed{\div}$ 45 $\boxed{=}$

Reverse Polish:

2 $\boxed{\Sigma}$ 8 $\boxed{\text{enter}}$ 2 $\boxed{\times}$ $\boxed{\Sigma}$ 21 $\boxed{\text{enter}}$ 3 $\boxed{\times}$
$\boxed{\Sigma}$ 13 $\boxed{\text{enter}}$ 4 $\boxed{\times}$ $\boxed{\Sigma}$ 6 $\boxed{\Sigma}$ $\boxed{\text{RCL}}$ 45 $\boxed{\div}$

(iv) Reverse Polish machine procedure without using $\boxed{\Sigma}$ is a little briefer:

Reverse Polish key strokes	*Display*
2 [enter]	2
8 [enter] 2 [×] [+]	18
21 [enter] 3 [×] [+]	81
13 [enter] 4 [×] [+]	133
6 [+]	139
45 [÷]	3·0888889

Very similar to the above problem is the following:

The percentages of the different stable isotopes in naturally occurring silicon are:

$$92 \cdot 2\% \ ^{28}Si, \quad 4 \cdot 7\% \ ^{29}Si, \quad 3 \cdot 1\% \ ^{30}Si$$

Calculate the weighted average atomic mass of naturally occurring silicon.

The necessary calculation to find the 'weighted average' is

$$\frac{\text{weighting} \times \text{value} + \text{weighting} \times \text{value} + \ldots}{\text{weighting} + \text{weighting} + \ldots}$$

i.e.

$$\frac{92 \cdot 2 \times 28 + 4 \cdot 7 \times 29 + 3 \cdot 1 \times 30}{92 \cdot 2 + 4 \cdot 7 + 3 \cdot 1}$$

The denominator here of course adds up to 100%. Using the pocket calculator the computed answer found is 28·109 so a reasonable value to quote would be 28·1. The 'weighting' is necessary to take account of the fact that the various atomic masses occur in different proportions.

Class intervals

Frequency tables are often presented with the data grouped in class intervals (e.g. 0 to 5, 6 to 10, . . .), rather than as specific individual values as the egg clutch sizes were in the earlier example (i.e. 1, 2, 3, . . .).

As an example:

A number of independent determinations of the normality of an acid solution were made with the following results:

Normality Recordings

x_i	0·092 to 0·094	over 0·094 to 0·096	over 0·096 to 0·098	over 0·098 to 0·101
Frequency f_i	12	16	6	3

Estimate the actual normality of the solution.

To do this the procedure is much as before but it is necessary to use the midpoints of the intervals. For example, there are 12 results in the 0·092 to 0·094 interval. We assume for the purposes of calculation that these 12 are equivalent to 12 at the midpoint of the interval, i.e. 12 at 0·093. The necessary calculation, then, is

$$\frac{\text{frequency} \times \text{midpoint} + \text{frequency} \times \text{midpoint} + \ldots}{\text{frequency} + \text{frequency} + \ldots}$$

i.e.

$$\frac{12 \times 0·093 + 16 \times 0·095 + 6 \times 0·097 + 3 \times 0·0995}{12 + 16 + 6 + 3}$$

[Note that the last interval is larger than the others.]

Using the calculator we get 0·09504054 so the normality is estimated to be 0·095.

GEOMETRIC MEAN

This less familiar average is useful (i) when the data are from some process growing or decaying exponentially rather than linearly, and (ii) for averaging relative changes.

The geometric mean of two numbers a, b is defined as

$$\sqrt{(a \times b)}$$

More generally the geometric mean of n numbers $a_1, a_2, a_3, \ldots, a_n$ is

$$\sqrt[n]{(a_1 \times a_2 \times \ldots \times a_n)}$$

Example

The population of the U.S.A. in 1820 was 9·6 millions and had grown to 17·1 millions by 1840. Estimate the population in 1830. Using a calculator for $\sqrt{(9·6 \times 17·1)}$ gives 12·812494 so the geometric mean estimate is 12·8 millions (the actual population in 1830 was 12·9 millions). Note that the arithmetic mean (13·4) is much less accurate.

HARMONIC MEAN

This is another type of average only occasionally come across.

The harmonic mean of $x_1, x_2, \ldots, x_n$ is defined as

$$\frac{n}{\left[\dfrac{1}{x_1} + \dfrac{1}{x_2} + \ldots + \dfrac{1}{x_n}\right]}$$

It can be useful in finding average speeds in problems such as: An athlete runs up a hill at 12 kmh^{-1} and down the hill again at 16 kmh^{-1}. What is his average speed? The arithmetic mean (14 kmh^{-1})

is inappropriate because the two rates are not maintained for equal amounts of time—the faster the athlete runs the less time he takes. The speeds are in fact maintained for times proportional to their reciprocals (inverse proportionality).

$$\text{Average speed} = 2/[\tfrac{1}{12} + \tfrac{1}{16}]$$

Using the calculator:

Algebraic: 12 ▢1/x ▢+ 16 ▢1/x ▢÷ 2 ▢x⇌y ▢=
or
Reverse Polish: 12 ▢1/x ▢enter 16 ▢1/x ▢+ 2 ▢x⇌y ▢÷

gives *13·714286.*

Some things to do

Exercise 7.1   Naturally occurring lead consists of approximately $1·37\%$ ^{204}Pb, $26·26\%$ ^{206}Pb, $20·82\%$ ^{207}Pb and $51·55\%$ ^{208}Pb. Find the atomic mass of lead.

Exercise 7.2   A batch of twelve resistors each marked 20 Ω are tested and found to have the following values (in ohms): 19·9, 20·0, 20·3, 20·2, 20·2, 20·0, 20·3, 20·2, 20·2, 20·1, 20·1, 20·0.
Find the mean resistance for the batch.

Exercise 7.3   The lengths of 126 cuckoo eggs were measured to the nearest 0·1 mm. The results are set out in the following frequency table. Find the mean egg length.

Interval	15·8 to 16·1	16·2 to 16·5	16·6 to 16·9
Frequency	10	27	39
Interval	17·0 to 17·3	17·4 to 17·7	17·8 to 18·1
Frequency	35	11	4

Exercise 7.4   A woman is murdered at 10 p.m. on a cold and windy night (0°C) and her body is discovered at 11 p.m. When a doctor measures the body temperature at midnight he finds it to be 18°C. Assuming the body temperature at the time of death to have been the normal 37°C, estimate the body temperature (*a*) when the body was found, (*b*) at 10.30 p.m.

Exercise 7.5   A jet plane flies from Heathrow to New York (J.F.K.), against a constant head-wind, at a ground speed of 403 knots. It returns later, this time flying down-wind, at a ground speed of 627 knots. What is its average ground speed for the round trip?

Exercise 7.6 A common use of the geometric mean is in *averaging relative changes*. An embryo's mass is at 1 week 200 g, at 2 weeks 400 g, and at 3 weeks 600 g. The relative growth in the period 1–2 weeks is $(400/200) \times 100 = 200\%$ and in the period 2–3 weeks is $(600/400) \times 100 = 150\%$. If we take the arithmetic average growth rate we get 175%. Is this sensible? Starting from 200 g and applying a 175% growth rate twice, we get $200 \xrightarrow{175\%} 350 \xrightarrow{175\%} 612 \cdot 5$ which does not give the 600 g result it should. Show that the *geometric* mean of the growth rates does give the correct final mass.

VARIANCE AND STANDARD DEVIATION CALCULATIONS FOR THE WHOLE POPULATION

The (arithmetic) mean, μ (Greek letter mu), of a set of numbers $x_1, x_2, \ldots, x_n$ is defined as

$$\mu = \frac{1}{n} \sum_{i=1}^{n} x_i = \frac{x_1 + x_2 + x_3 \ldots + x_n}{n}$$

This is an exact result. Because we are using all the numbers in the population (as it is called) we are not taking a sample and *estimating* (that is discussed in a later section).

The mean is a useful measure but does not indicate how widespread are the values which contributed to its value. Hence the need for what is called the *variance*, σ^2 (sigma squared), of a set of numbers, and its square root, the standard deviation, σ.

Variance is defined as

$$\sigma^2 = \frac{1}{n} \sum_{i=1}^{n} (x_i - \mu)^2$$

In other words:

To find the variance take each of the n numbers in turn, subtract the mean, square the result, add up all the squared values and divide by n.

Actually there is an alternative way to compute the variance:

$$\sigma^2 = \frac{1}{n} \sum_{i=1}^{n} x_i^2 - (\mu)^2$$

In other words: add up the squares of all the n numbers, divide by n and subtract the square of the mean. The advantages of this second formula are that it is easier to use and allows one to calculate the mean at the same time rather than needing to find it beforehand. It therefore requires many less key strokes.

Example:

Find the variance and standard deviation of 6, 2, 7, 9, 8.

	Algebraic key strokes	Display	Reverse Polish key strokes	
(a)	6 [STO] [x²] [+]	36	6 [STO] [x²] [enter]	(a)
(b)	2 [M+] [x²] [+]	40	2 [Σ] [x²] [+]	(b)
(c)	7 [M+] [x²] [+]	89	7 [Σ] [x²] [+]	(c)
(d)	9 [M+] [x²] [+]	170	9 [Σ] [x²] [+]	(d)
(e)	8 [M+] [x²] [=]	234	8 [Σ] [x²] [+]	(e)

Note: on some machines [Σ] must be followed by [last x] to recover the data value to be squared.

Now we have Σx_i^2 ($=234$) in the display and Σx_i ($=32$) in the memory. We want

$$\frac{\Sigma x_i^2}{5} - \left(\frac{\Sigma x_i}{5}\right)^2$$

If the calculator has the facility for interchanging the display and the memory ([X⇌M] key) or if more than one memory is available then it is a simple matter to finish off the calculation. Otherwise a little more ingenuity is called for (or pencil and paper!)

To get the variance we can calculate

$$(5 \times \Sigma x_i^2 - (\Sigma x_i)^2)/5^2$$

	Algebraic key strokes	Display	Comment	Reverse Polish key strokes	
(f)	[×] 5 [−]	1170·	$5\Sigma x_i^2$	5 [×]	(f)
(g)	[RCL] [x²]	1024·	$(\Sigma x_i)^2$	[RCL] [x²]	(g)
(h)	[÷] 25 [=]	5·84	The variance of the population, σ^2.	[−] 25 [÷]	(h)
(i)	[√x]	2·4166092	The standard deviation of the population, σ.	[√x]	(i)

With Reverse Polish machines it is also possible to use the rolling stack in conjunction with one memory. This is left to the interested reader to investigate. It turns out that the variance is of much theoretical importance but the standard deviation is easier to understand—being some sort of 'average' deviation from the mean.

General formula for variance calculation on a pocket calculator.

$$\sigma^2 = \frac{n\Sigma x_i^2 - (\Sigma x_i)^2}{n^2}$$

Some things to do

Exercise 7.7 The geiger counter readings taken in successive intervals when testing a sample of uranium ore were as follows: 123, 131, 118, 120, 119, 130, 128, 127, 124, 126, 130, 127, 122. Calculate the mean, variance and standard deviation.

Exercise 7.8 When data are grouped the mean is found from $\Sigma f_i x_i / \Sigma f_i$ where f_i is the frequency for x_i. This is effectively what we did for the Blackbird eggs example (page 121). If the data are presented in class intervals then x_i is the middle value of the ith class interval. This we did for the acid normality example (page 122).

In a similar way the variance can be found for grouped data. The formula is

$$\sigma^2 = \frac{\Sigma f_i x_i^2}{\Sigma f_i} - \left(\frac{\Sigma f_i x_i}{\Sigma f_i}\right)^2$$

Find the variance of the cuckoo egg sizes of Exercise 7.3.

MEAN, VARIANCE AND STANDARD DEVIATION USING SAMPLES

It is usual in statistical work for the population size to be very large and only *samples* can be analysed. Having obtained a sample it is possible to compute the mean and variance *for the sample* but will they be representative of the whole population? Two main issues arise. Firstly, what are the best estimates of the statistical measures for the whole population? Secondly, how reliable are those measures? (To be dealt with in a separate section.) It turns out, as one might expect, that the mean of the sample, $\bar{x}$, is the best estimate of the mean of the total population, μ. However, for the *variance* it happens that the sample variance s^2 *underestimates* the population variance σ^2. For a sample of size n the best estimate of σ^2 is not s^2 but $(n/(n-1))s^2$.

This means that the formula for estimating the variance σ^2 using a sample must be:

$$\sigma_{est}^2 = \frac{\Sigma(x_i - \bar{x})^2}{n-1} \quad \text{and } not \quad \frac{\Sigma(x_i - \bar{x})^2}{n}$$

Example

A batch of twelve 60 W light bulbs is taken at random from a consignment of 10000 bulbs and burned to find their life. The results (in hours) are 983, 1009, 1005, 1021, 990, 1007, 1040, 1012, 994, 1021, 1003, 1011. Estimate the mean, variance and standard deviation of lamp life for the whole consignment.

A good version of the formula to use on a pocket calculator is

$$\sigma_{est}^2 = \frac{n\Sigma x_i^2 - (\Sigma x_i)^2}{n(n-1)}$$

which is very similar to that given for σ^2 earlier (page 126).

Using a very similar procedure to that for σ^2 we have, working in scientific notation on the display:

Algebraic key strokes	Display	Comment	Reverse Polish key strokes
(a) 983 STO			983 STO
x^2 +	9·6629 05	x_1^2	x^2
(b) 1009 M+			1009 Σ x^2
x^2 +	1·9844 06	$x_1^2 + x_2^2$	+
. . . . .	. . . .	. . . .	. . . . .
(l) 1011 M+			1011 Σ x^2
x^2 =	1·2195 07	Σx_i^2	+
			[Note Σ may need last x to follow it.]

We now have Σx_i^2 in the display and Σx_i in the memory.

Algebraic key strokes	Display	Comment	Reverse Polish key strokes
(m) × 12 −	1·4634 08	$12\Sigma x_i^2$	12 ×
(n) RCL x^2			RCL x^2
=	3·0816 04	$12\Sigma x_i^2 - (\Sigma x_i)^2$	−
(o) ÷ 12 ÷			12 ÷ 11
11 =	2·3345 02	variance σ_{est}^2	÷
(p) $\sqrt{x}$	1·5279 01	std. devn. σ_{est}	$\sqrt{x}$
(q) RCL ÷ 12	1·008 03	Mean μ_{est}	RCL 12 ÷
=			

Thus mean $\approx 1008\cdot0$, variance $\approx 233\cdot5$, std. devn. $\approx 15\cdot28$.

Of course the difference between using n and $n-1$ in the variance formula is insignificant for large n. It is only of importance when small samples are being used. Some pocket calculators will automatically calculate means and standard deviations and for the latter $n-1$ will generally be used. So if you wish to compute s rather than σ_{est}—in other words you have got the whole population as your sample, and it is small—then you must take the result given, σ_{est}, and multiply it by $\sqrt{((n-1)/n)}$. Some machines offer both formulae.

THE NORMAL DISTRIBUTION (GAUSSIAN DISTRIBUTION)

Many physical characteristics are distributed in a way which closely conforms to the Normal curve (fig. 7.1). Most values cluster about the mean and only a few deviate far from the mean. The Normal distribution curve is a mathematically defined one and it is known that about 68% of all the area lies within 1 standard deviation of the mean, 95% within 2 standard deviations and 99·7% within 3 standard deviations. The height of adult males in the population of a country closely follow the Normal curve of distribution. If the mean is 175 cm and the standard deviation 9 cm, then 68% of the adult males will be between 175−9 cm and 175+9 cm, i.e. 166 cm to 184 cm, 16% will be taller than 184 cm, 16% will be shorter than 166 cm.

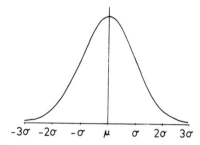

-3σ -2σ $-\sigma$ μ σ 2σ 3σ

Fig. 7.1   Normal curve of distribution.

It is found that when a number of samples, each of size n, are taken from a population the sample means themselves often conform to a Normal distribution pattern, clustering about the mean of the whole population from which the samples are drawn. This is true even in cases where the population itself is not Normally distributed, provided n is large enough (> 30 say).

The standard deviation of these sample means, σ_m, is related to the standard deviation of the population, σ: $\sigma_m = \sigma/\sqrt{n}$ where n is the size of each sample.

This value σ_m is known as the Standard Error of the mean. Clearly the larger the sample size n the smaller will be σ_m, the standard error (or S.E.).

The size of the S.E. is an indication of how reliable the mean estimate is. The estimated mean (of the population) will be:

within $\pm 1\sigma_m$ of the true mean with probability 68%

within $\pm 2\sigma_m$ of the true mean with probability 95%

within $\pm 3\sigma_m$ of the true mean with probability 99·7%

Example

The light bulb example already discussed was based on a sample of size $n = 12$. The calculated mean was $\mu_{est} = 1008 \cdot 0$, and $\sigma_{est} = 15 \cdot 28$.

The standard error of the mean is $\sigma/\sqrt{12}$. However we do not actually know the true σ for the whole consignment and to find it would destroy all the bulbs! σ_{est} is our best estimate of it so that we must use:

$$\text{S.E. } (\sigma_m) \approx \frac{\sigma_{est}}{\sqrt{12}} \approx \frac{15 \cdot 28}{\sqrt{12}} \approx 4 \cdot 41$$

95% of all samples will lie within 2 standard errors of the mean, i.e. within approximately $\pm 2 \times 4 \cdot 41$. So our 95% confidence limits are $1008 \cdot 0 \pm 8 \cdot 82$ which are $999 \cdot 18$ and $1016 \cdot 82$. We can be 95% confident that the average bulb life for the whole consignment (i.e. the true mean) lies in the range $999 \cdot 18$ to $1016 \cdot 82$ (fig. 7.2).

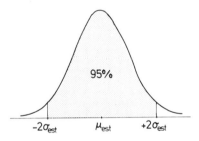

Fig. 7.2   Area within two standard deviations of the mean, for the Normal distribution.

For even greater certainty we might prefer to work to 3 standard errors, giving us 99.7% confidence limits of $1008 \cdot 0 \pm 3 \times 4 \cdot 41$, i.e. $994 \cdot 77$ to $1021 \cdot 23$.

[*Note:* it would be more appropriate here to use the *t*-distribution since the sample size is appreciably smaller than 25. However the *t*-distribution is not discussed in this book.]

Some things to do

Exercise 7.9   Certain meat pies should contain at least 25% meat in order to conform to the Trade Descriptions Act. Following a complaint about one brand, a Food Inspector analyses ten pies bought at different times from different shops. His analysis findings are: $26 \cdot 3$, $27 \cdot 2$, $25 \cdot 4$, $25 \cdot 2$, $25 \cdot 3$, $27 \cdot 0$, $25 \cdot 7$, $25 \cdot 5$, $25 \cdot 9$, $25 \cdot 4$.

Find the mean and standard deviation of the sample. Estimate the mean and standard deviation of all the meat pies of this type in the shops. The Inspector can be 95% confident that the estimated mean will be within 2 standard errors of the true mean. Find the confidence limits. What would the confidence limits be for 99·7% certainty (i.e. ± 3 standard errors)?

Exercise 7.10 Chromatographic tests are made on six samples drawn from a canister of helium gas supposedly free of nitrogen. The levels of nitrogen found, in parts per million, are: 15·3, 14·4, 13·9, 16·2, 13·2, 13·7.

Find the best estimate of the nitrogen level in the canister of helium. What limits can be placed on the actual nitrogen level with a confidence of 95% (i.e. ± 2 standard errors)?

CORRELATION: PEARSON'S r

There are various measures of correlation—which seek to measure how closely two sets of data are related (e.g. I.Q. and parental income).

The formula for Pearson's coefficient of correlation, r, is

$$r = \frac{Sxy}{\sqrt{(Sxx)}\sqrt{(Syy)}}$$

where

$$Sxy = \Sigma x_i y_i - \frac{\Sigma x_i \Sigma y_i}{n}$$

$$Sxx = \Sigma x_i^2 - \frac{(\Sigma x_i)^2}{n}$$

$$Syy = \Sigma y_i^2 - \frac{(\Sigma y_i)^2}{n}$$

where the two sets of data are

$$x_1, x_2, x_3, \ldots x_i, \ldots x_n$$

$$y_1, y_2, y_3, \ldots y_i, \ldots y_n$$

In fact the Sxx and Syy terms are simply n times the variances for the x and y populations and we have already discussed their calculation. It will be seen that both Σx_i and Σy_i are needed in finding Sxy so it is best to first calculate Sxx and Syy noting the values of Σx_i and Σy_i for subsequent use.

r can range from -1 to $+1$.

$r = 0$ indicates no correlation between the x and y values.

$r = 1$ shows perfect agreement—as x increases so does y proportionately without exception.

$r = -1$ shows a perfect inverse relationship—as x increases y decreases proportionately without exception.

Example

In an experiment into the growing of winter wheat the mean soil temperature at a soil depth of 8 cm and the days needed for germination were recorded:

Soil temperature °C	5	5·5	6	6·5	7	7·5	8	8·5
Days for germination	40	36	32	27	23	19	19	20

Find the correlation between the soil temperature and the germination time.

Using the same key-stroke sequence as for variance (pages 125–126), which we leave as an exercise, we obtain

$$Sxx = 10·5$$
$$Syy = 468$$

During the above computations we note that

$$\Sigma x_i = 54$$
$$\Sigma y_i = 216$$

We can now proceed to calculate $\Sigma x_i y_i$ and then Sxy and finally find r itself. The key-stroke details are given on page 133. Thus $r \approx -0·96$ indicating a very high inverse correlation between the variables as is obvious from studying the data.

CORRELATION: SPEARMAN'S ρ

Another correlation coefficient is Spearman's ρ (Greek letter rho) which compares ranks (i.e. relative positions) rather than actual data values. The formula is

$$\rho = 1 - \frac{6\Sigma d^2}{n(n-1)(n+1)}$$

where Σd^2 is the sum of the squares of the difference between the corresponding ranks.

As for Pearson's r, Spearman's ρ lies between -1 and $+1$ and values have similar significance. Spearman's coefficient measures the extent to which the relationship is continuously increasing or continuously decreasing. Pearson's coefficient measures the extent to which the relationship increases or decreases *as a straight line* (linearly).

Algebraic key strokes	Display	Comment	Reverse Polish key strokes (using memory)	Reverse Polish key strokes (using stack)	Display for Reverse Polish key strokes using stack
5 [×] 40 [=] [STO]	200·	$x_1 y_1$	5 [enter] 40 [×] [Σ]	5 [enter] 40 [×]	200·
5·5 [×] 36 [=] [M+]	198·	$x_2 y_2$	5·5 [enter] 36 [×] [Σ]	5·5 [enter] 36 [×] [+]	398·
. . . . . .	. . . .	. . . .	. . . . .	. . . . .	. . . .
8·5 [×] 20 [=] [M+]	170·	$x_8 x_8$	8·5 [enter] 20 [×] [Σ]	8·5 [enter] 20 [×] [+]	1391·
54 [×] 216 [÷] 8 [−]	1458·	$\dfrac{\Sigma x_i \Sigma y_i}{8}$	54 [enter] 216 [×] 8 [÷]	54 [enter] 216 [×] 8 [÷]	1458·
[RCL] [=] [+/−]	−67·	S_{xy}	[RCL] [−] [+/−]	[−]	−67·
[÷] 10·5 [√x] [÷] 468 [√x] [=]	−·95577843	r	10·5 [√x] [÷] 468 [√x] [÷]	10·5 [√x] [÷] 468 [√x] [÷]	−·95577843

Example

A keen lecturer believes that the more lectures a student attends the better is his or her examination performance. He keeps a record of term-time attendances and compares these with the final examination grades for a science course:

Attendances	50	48	47	43	41	39	38	36	35	26
Exam. performance (%)	72	37	60	60	29	80	50	75	26	40

First we convert these to rank order form, largest numbers first.

Attendances	1	2	3	4	5	6	7	8	9	10
Exam. performances	3	8	$4\frac{1}{2}$	$4\frac{1}{2}$	9	1	6	2	10	7

[Note that 4th and 5th examination places are shared so each is allocated $4\frac{1}{2}$.] We can now proceed to calculate:

Algebraic key strokes	*Display*	*Comment*	*Reverse Polish key strokes*
1 $-$ 3 $=$ x^2 $\boxed{\text{STO}}$	4·	d_1^2	1 $\boxed{\text{enter}}$ 3 $-$ x^2 $\boxed{\text{STO}}$
2 $-$ 8 $=$ x^2 $\boxed{\text{M+}}$	36·	d_2^2	2 $\boxed{\text{enter}}$ 8 $-$ x^2 $\boxed{\text{M+}}$
.....	...	...	......
10 $-$ 7 $=$ x^2 $\boxed{\text{M+}}$	9·	d_{10}^2	10 $\boxed{\text{enter}}$ 7 $-$ x^2 $\boxed{\text{M+}}$
$\boxed{\text{RCL}}$	130·5	Σd^2	$\boxed{\text{RCL}}$
$\boxed{+/-}$ $\times$ 6 $\div$ 10 $\div$	$-78\cdot3$	$-6\Sigma d^2/n$	$\boxed{+/-}$ 6 $\times$ 10 $\div$
9 $\div$ 11 $+$ 1 $=$	·20909091	ρ	9 $\div$ 11 $\div$ 1 $+$

Thus there is a very small positive correlation between attendance and marks ($\rho \approx 0.21$) but not a strong relationship at all. The lecturer is disappointed.

LINEAR REGRESSION

A regression equation is a formula connecting two variables which are to some extent correlated. If two variables are strongly correlated we may wish to predict the value of one of the variables associated with a particular value of the other variable. In *linear regression* we try to fit a straight line through the set of points (x_i, y_i). If we want to predict y from x we usually choose the line that minimizes the sum of squares of the differences between the actual y_i and the values predicted by the straight line equation (fig. 7.3).

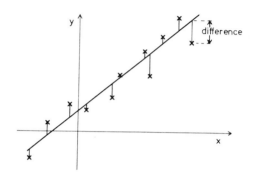

Fig. 7.3   The 'best' straight line through a set of points.

For two related sets of data $x_1 \ldots x_n$ and $y_1 \ldots y_n$ in order to find the regression equation to predict y given x, three values must be computed:

(i) $\bar{x}$; (ii) $\bar{y}$; (iii) $b_{yx} = \dfrac{Sxy}{Sxx}$

Now the means $\bar{x}$ and $\bar{y}$ are well understood and readily calculated, and b_{yx} is called the regression coefficient of y on x.

As before

$$Sxy = \Sigma x_i y_i - \frac{\Sigma x_i \Sigma y_i}{n}$$

$$Sxx = \Sigma x_i^2 - \frac{(\Sigma x_i)^2}{n}$$

and their calculation has been discussed earlier in the section on correlation (Pearson's r). The regression equation of 'y on x' is then:

$$y - \bar{y} = b_{yx}(x - \bar{x})$$

or in slope-intercept form

$$y = b_{yx}x + [\bar{y} - b_{yx}\bar{x}]$$

For any particular value of x this gives the best estimate of y.

Example

Scientists investigating the control of locusts by spraying, conduct experiments in which batches of 100 locusts are sprayed with varying concentrations of a toxic chemical. The deaths within two days are recorded:

x: Concentration (parts per 1000)	1	2	4	6	10	15	20
y: Deaths per 100 locusts	21	18	33	50	76	90	96

Find the regression equation of deaths on concentration. What % of deaths can be expected using 5 parts per 1000, and 10 parts per 1000?

We have to calculate $\bar{x}$, $\bar{y}$, Sxy, Sxx, b_{yx}. It should be noted that the calculation of Sxx will incidentally produce Σx_i so we can easily find $\bar{x} = \Sigma x_i / n$ from that. Σx_i and Σy_i are both needed for Sxy. The best procedure is therefore to compute the statistics in the order

$$Sxx, \bar{x}, \bar{y}, Sxy, b_{yx}$$

(i) Sxx This computation procedure has been described for correlation (Pearson's r). In our example $Sxx = 301 \cdot 42857$

(ii) $\bar{x}$ $\Sigma x_i = 58$ and so $\bar{x} = 8 \cdot 2857143$

(iii) $\bar{y}$ $\Sigma y_i = 384$ so $\bar{y} = 54 \cdot 857143$

(iv) Sxy This computation procedure has been described for correlation (Pearson's r). In our example $Sxy = 2849 \cdot 4694$

(v) $b_{yx} = \dfrac{Sxy}{Sxx} = 9 \cdot 453216$

Thus the regression equation of y on x is approximately:

$$y - 54 \cdot 86 = 9 \cdot 45(x - 8 \cdot 28)$$

or in slope-intercept form

$$y = 9 \cdot 45x - 23 \cdot 3.$$

When concentration (x) is 5 the estimated deaths (y) will be given by

$$y = 54 \cdot 86 + 9 \cdot 45(5 - 8 \cdot 28) = 23 \cdot 90$$

For concentration $x = 10$

$$y = 54 \cdot 86 + 9 \cdot 45(10 - 8 \cdot 28) = 71 \cdot 15$$

So we can expect about 24% deaths using 5 p.p. thousand and 71% deaths using 10 p.p. thousand.

Some things to do

Exercise 7.11   Karl Pearson studied the inheritance of height and other physical attributes and published many interesting tables, with Alice Lee, in 1903. The table below is a much reduced version of one such table which shows the 'regression towards the mean', i.e. extremely short or tall parents tend to have offspring with rather less extreme heights:

Father's height (inches)	Son's height (inches)	Father's height (inches)	Son's height (inches)
x	y	x	y
63	65	68	70
63	66	69	66
64	67	69	70
65	66	69	69
65	64	70	66
66	67	70	68
66	68	70	70
66	68	71	69
67	68	72	70
67	67	74	72
67	69	74	71
68	68	75	72

Find the regression line for y on x. How linear is the correlation between fathers' and sons' heights?

Exercise 7.12 Seven sixth formers were examined in physics and chemistry with the following results:

Physics	36	52	17	51	37	38	69
Chemistry	51	73	34	65	53	60	78

Find the Spearman's ρ coefficient and Pearson's r coefficient.

Exercise 7.13 M. R. Scott (1973) has presented the results of two very similar Final degree mathematics examination papers sat by a group of 19 students:

Candidate	Paper 1	Paper 2
A	75	72
B	66	72
C	83	71
D	87	48
E	97	63
F	49	45

(*continued*)	Candidate	Paper 1	Paper 2
	G	94	100
	H	80	73
	I	100	71
	J	33	38
	K	37	43
	L	38	37
	M	44	51
	N	44	40
	O	49	34
	P	42	34
	Q	55	45
	R	77	70
	S	75	63

Find the mean and standard deviation for each paper, and Pearson's coefficient for correlation for the two papers.

THE χ^2 TEST (CHI-SQUARED TEST)

This statistical method tests whether observed results differ *significantly* from what might be expected to occur. For example, if a coin is tossed 100 times and it falls: Heads 60 times, Tails 40 times, should we conclude that the coin is biased, or is the result reasonable for an unbiased coin?

The approach used is to draw up a table showing the observed (O) and expected (E) results:

	O	E
Heads	60	50
Tails	40	50

For *each* of the variables (Heads and Tails) we calculate $(O-E)^2/E$ and add together the results ($\Sigma[(O-E)^2/E]$). This gives us a χ^2 *statistic*, which is a measure of the discrepancy between the expected and observed results. The bigger the value of the χ^2 statistic the bigger is the discrepancy. In our example:

$$\chi^2 = \frac{(60-50)^2}{50} + \frac{(40-50)^2}{50} = 4\cdot0$$

With one further piece of information, the 'degrees of freedom', we can look up the value $4\cdot0$ in a χ^2 table which will indicate how probable is the observed discrepancy. The interested reader should consult a Statistics book for details about degrees of freedom. In this example there are two variables, Heads and Tails, but if one is specified (e.g. 60 Heads) then there is no freedom of choice for the other (it must be 40 Tails). There is only 1 degree of freedom here (only one variable can be freely chosen). Looking up the appropriate χ^2 table we find that for 1 degree of freedom a χ^2 value of 3·84 or bigger has a probability of 5% of occurring, so our value $\chi^2 = 4$, which is more than the 5% critical value (3·84), would arise in less than 1 in 20 such coin-tossing experiments if the coin is unbiased. We might conclude therefore that there are good grounds for suspecting that the coin is biased *at the 5% level*. However, if we wished to work at the 1% level (i.e. give the coin the benefit of the doubt unless the probability of a discrepancy as large or larger than that found is under 1%) the critical χ^2 value is 6·64. The computed χ^2 statistic, being 4·0, is less than that and so the discrepancy is not significant *at the 1% level*.

Some things to do

Exercise 7.14 F. A. Bollen (1972) investigated the relationship between College of Education students' attitudes towards the discovery approach to Science Teaching and the number of years of science teaching which the students had previously received. The table of results was as follows:

| | Years of previous study | | | | | |
	0–6		7–10		11–20	
Attitude Score	O	E	O	E	O	E
39–46	18	14·5	13	12·8	7	10·7
47–50	10	15·3	15	13·5	15	11·2
51–58	13	11·2	8	9·7	8	8·1

degree of freedom: 4

Bollen surmised that there was no connection between attitude and previous experience.

Calculate the value of χ^2 for this table and compare with the 5% critical value of 9·5.

(Hint: $\chi^2 = \sum \dfrac{(O-E)^2}{E} = \dfrac{(18-14·5)^2}{14·5} + \ldots + \dfrac{(8-8·1)^2}{8·1}$ (nine terms)).

Exercise 7.15 It is known that ethylene is evolved by plants and that it affects plant metabolism in various ways. A biologist sets up an experiment to determine whether ethylene has any effect upon the development of fruit in certain flowering plants.

(a) Firstly he selects 40 cut bluebell plants (monocotyledonous) and places each in a jar containing 2% sucrose solution. Each plant is covered with a plastic bag and all the jars are placed for 10 days in a darkened room. Then he analyses the sizes of fruits on each plant, using two categories for the fruits: 'large' and 'small'. He notes the percentage of large fruits each plant has and tabulates his observed results—as in Column 1 of Table A. Having dealt with the control group he next repeats the experiment with 40 more bluebell plants, but this time adds 5 drops of an ethylene-producing solute (technical ethephon) to the sucrose solution in each jar. His observed results for the experimental group are shown in Column 3 of Table A.

TABLE A   *Bluebell*

	1	2	3	4
Percentage of large fruits	Control	Expected value for control	Experimental	Expected value for experimental
Over 60%	5	8	11	8
41%–60%	10	13·5	17	13·5
21%–40%	14	11·5	9	11·5
0%–20%	11	7	3	7
	40	40	40	40

Now is there any significant difference between the control and experimental group results? Has the ethylene-producing solute produced an effect not reasonably attributable to random variation? To test this we must calculate the expected values for each category for both groups. As the two groups are equal in size (40 each) these are simply the averages of the two numbers in each category—as shown in Columns 2 and 4. Now we must calculate the χ^2 statistic for the eight observed results. Does it exceed the critical value of 7·8 (5% level, 3 degrees of freedom). What is the conclusion?

(b) The biologist then tests a dicotyledonous plant—Stitchwort—and his findings are shown in Table B. Are these results significant?

TABLE B   *Stitchwort*

	1	2	3	4
Percentage of large fruits	Control	Expected value for control	Experimental	Expected value for experimental
Over 60%	21	. . .	12	. . .
41%–60%	25	. . .	12	. . .
21%–40%	38	. . .	9	. . .
0%–20%	16	. . .	6	. . .
	100	100	39	39

(Hint: To find the expected values proceed as follows:

In the 'over 60%' category there are 33 plants. Assuming there is no real difference between control and experimental groups, these 33 should be proportionately distributed as $(100/139) \times 33$ in the Control Group (Col. 1) and $(39/139) \times 33$ in the Experimental Group (Col. 3). Thus for the 'over 60%' category the expected values are 23·74 for the Control Group and 9·26 for the Experimental Group.) [P. W. Freeland (1976) described a similar experiment but the statistical analysis has been amended here.]

Exercise 7.16   In an experiment to investigate whether the sizes of woodlice colonies and the humidity of the habitat are related, a zoology student makes a study of 284 locations. He classifies the woodlice colonies into four sizes and categorizes the humidity into five levels, testing by means of cobalt thiocyanate paper. His results are:

Humidity Population	(Low) 1	2	3	(High) 4	5	Row Totals
(Small) A	7	22	20	6	3	58
B	7	16	31	13	14	81
C	6	9	28	41	13	97
(Large) D	5	7	6	12	18	48
Column Totals	25	54	85	72	48	284

Has the student disproved the hypothesis that no relationship exists? (Degrees of freedom: 12; 5% critical value: 21·03, 1% critical value: 26·22.)

STATISTICAL METHODS AND THE POCKET CALCULATOR

This book is not the place in which to go into the detailed explanation of statistical analysis any further. It should be clear that a pocket calculator is a valuable tool for both minor and extensive calculation problems which arise in statistical work. Indeed calculators are increasingly coming onto the market with built-in statistical functions. Means and standard deviation are now commonplace. Some machines have a key which adds to the memory the square of the number in the display. Usually it is marked $\boxed{M+x^2}$ to distinguish it from the key which adds the number displayed into the memory which is designated by $\boxed{M+}$ or $\boxed{\Sigma}$ or possibly $\boxed{M+x}$. This $\boxed{M+x^2}$ key is obviously very useful when computing variance, standard deviation, correlation, linear regression, etc. Indeed there are calculators available which automatically calculate correlation coefficients, and perform linear (and multiple) regression analyses, and calculate χ^2.

At least one calculator on the market will provide values for various probability distributions—Binomial, Poisson, Normal (or Gaussian) which we shall discuss in Chapter 8.

When the realm of programmable calculators is entered (see Chapter 16) there seems to be no limit, with analysis of variance and a host of other packages available.

8. Probability

Some of the more sophisticated calculators have a key marked $\boxed{n!}$ or $\boxed{x!}$. Given a positive integer, k, in the display, depressing $\boxed{n!}$ produces the value of $1 \times 2 \times 3 \times \ldots \times (k-1) \times k$ which is $k!$ ('k factorial').

Example

$$6! \text{ is } 6 \times 5 \times 4 \times 3 \times 2 \times 1 = 720$$

so 6 $\boxed{n!}$ will give 720.

Successive factorial numbers get very large very quickly: $10! = 3628800$ for example. The largest factorial which most calculators can compute is $69! \approx 1.71 \times 10^{98}$ because $70! \approx 1.2 \times 10^{100}$ and only two digits are ever allowed for the exponent in scientific notation on a calculator. [The Corvus 500 can calculate up to 120! but any hundreds digit in the exponent is not displayed].

For the sake of completeness the definition $0! = 1$ is made. Factorials occur in the theory of arrangements (permutations and combinations) and in many branches of mathematics.

There is a very useful approximation to $n!$ known as Stirling's formula, first given in 1718:

$$n! \approx \sqrt{(2\pi)} n^{n+\frac{1}{2}} e^{-n} \left(1 + \frac{1}{12n}\right)$$

Example

(i) $5! = 120$ and by Stirling's formula we get 119·98615

(ii) $50! \approx 3.041409 \times 10^{64}$ using a calculator with $\boxed{n!}$ and
$50! \approx 3.041405 \times 10^{64}$ using Stirling's formula.

The $[1 + (1/12n)]$ factor is not very important, being so close to 1. Using the simplified Stirling's formula $n! \approx \sqrt{(2\pi)} n^{n+\frac{1}{2}} e^{-n}$ we get

(i) $5! \approx 118.01917$

(ii) $50! \approx 3.0363446 \times 10^{64}$

which are still quite accurate values.

PERMUTATIONS

The number of ways in which n different objects can be arranged in order is $n!$ This is called the number of *permutations* of n objects.

If only r of the n objects are selected, the number of ways of ordering them is $n!/(n-r)!$ and this is the 'number of permutations of r from n' given the symbol ^nP_r.

Example 1

How many ways can the letters A, B, C, . . . X, Y, Z be ordered taking 12 at a time?

$$^{26}P_{12} = \frac{26!}{(26-12)!} = \frac{26!}{14!} \approx 4 \cdot 6261 \times 10^{15}$$

Key strokes might be:

Algebraic: 26 $\boxed{n!}$ $\boxed{\div}$ 14 $\boxed{n!}$ $\boxed{=}$

Reverse Polish: 26 $\boxed{n!}$ 14 $\boxed{n!}$ $\boxed{\div}$

although on some machines $\boxed{n!}$ obliterates any previous intermediate calculations, in which case the memory must be used and each $\boxed{n!}$ calculated separately.

Example 2

How many permutations of 100 different objects taking 30 at a time are there?

Now this cannot directly be calculated on a machine because $^{100}P_{30} = 100!/70!$ and neither $100!$ nor $70!$ can be evaluated, although the answer is not itself too large. We can write $100!/70! = 100 \times 99 \times \ldots \times 71$ (cancelling out all the other factors) but it would be quite an effort to calculate that lot!

Stirling's formula provides an alternative method if some care is taken to work on both denominator and numerator together so that the numbers do not exceed the calculator's capacity.

$$\frac{100!}{70!} \approx \frac{\sqrt{(2\pi)} \times 100^{100 \cdot 5} \times e^{-100}}{\sqrt{(2\pi)} \times 70^{70 \cdot 5} \times e^{-70}}$$

$$\approx \frac{\sqrt{(2\pi)} \times 100^{70 \cdot 5} \times 100^{30} \times e^{-70} \times e^{-30}}{\sqrt{(2\pi)} \times 70^{70 \cdot 5} \times e^{-70}}$$

$$\approx \left(\frac{100}{70}\right)^{70 \cdot 5} \times \left(\frac{100}{e}\right)^{30}$$

Note that we could not evaluate 100^{100} or 70^{70} on a calculator but $(100/70)^{70 \cdot 5}$ and $(100/e)^{30}$ are quite manageable numbers.

Algebraic key strokes	*Display*	*Comment*	*Reverse Polish key strokes*
100 $\boxed{\div}$ 70 $\boxed{=}$	1·4286 00	100/70	100 $\boxed{\text{enter}}$ 70 $\boxed{\div}$
$\boxed{y^x}$ 70·5 $\boxed{=}$ $\boxed{\text{STO}}$	8·3289 10	$(100/70)^{70\cdot5}$	70·5 $\boxed{y^x}$
100 $\boxed{\div}$ 1 $\boxed{e^x}$			100 $\boxed{\text{enter}}$ 1 $\boxed{e^x}$
$\boxed{=}$	3·6788 01	$(100/e)$	$\boxed{\div}$
$\boxed{y^x}$ 30 $\boxed{=}$	9·3576 46	$(100/e)^{30}$	30 $\boxed{y^x}$
$\boxed{\times}$ $\boxed{\text{RCL}}$ $\boxed{=}$	7·7939 57	answer	$\boxed{\times}$

Cases when objects not all different

The number of permutations of n objects of which p are of one sort, q of a second sort, r of a third sort . . . is $n!/(p!q!r!\ldots)$

Example

How many distinct permutations of the letters of SCIENTISTS are possible? $n=10$, $p=3$ (letter S), $q=2$ (letter I), $r=2$ (letter T), therefore the number of ways is $10!/(3!2!2!)=151200$.

Cases when repetitions are allowed

If an object is selected from n different objects, replaced, then a new object is selected, and so on for a total of r selections, the number of permutations possible is n^r.

Example

A chimpanzee sits at a simplified typewriter and types six letters. How many possible outcomes are there?

Each of the six letters can be any one of the 26 of the alphabet so the number of outcomes is

$$26 \times 26 \times 26 \times 26 \times 26 \times 26 = 26^6 \approx 3\cdot089 \times 10^8$$

(using the $\boxed{y^x}$ key).

COMBINATIONS

The number of ways of selecting r objects from n different objects, no regard being made to order, is $n!/[(n-r)!r!]$ given the symbol ^nC_r. This is said to be the number of *combinations* of r objects from n.

Example 1

How many 3 letter combinations can be made from the letters of CHEMISTRY?

$$^9C_3 = \frac{9!}{6!3!} = 84$$

Example 2

A biologist wishes to use 75 rats in an experiment and has 200 to choose from. How many different ways can he select his rats?

$$^{200}C_{75} = \frac{200!}{75!125!}$$

Again recourse to Stirling's formula is made:

$$\frac{200!}{75!125!} \approx \frac{\sqrt{(2\pi)} \times 200^{200 \cdot 5} \times e^{-200}}{\sqrt{(2\pi)} \times 75^{75 \cdot 5} \times e^{-75} \times \sqrt{(2\pi)} \times 125^{125 \cdot 5} \times e^{-125}}$$

$$\approx \left(\frac{200}{75}\right)^{75 \cdot 5} \times \left(\frac{200}{125}\right)^{125} \times \frac{1}{125^{0 \cdot 5} \times \sqrt{(2\pi)}}$$

$$\approx 1 \cdot 69 \times 10^{56}$$

Some things to do

Exercise 8.1   A student has 9 different books on his shelf. In order to put off getting down to studying he decides to see in how many different ways he can order them on his shelf. If he takes 5 seconds to make each arrangement, how long will it be before he resumes his studies?

Exercise 8.2   It is estimated that there are 2 million species of insect in the world. If each is to be given a unique code consisting of r letters from the alphabet how many letters must be in the code?

Exercise 8.3   A biochemist has 8 solutions which he has prepared as possible cures for rabies. He decides to try taking combinations of the solutions and injecting each different mixture into one rabid dog. (*a*) He decides to take the solutions 3 at a time. How many dogs does he need for the experiment? (*b*) He then decides to try out all other possible combinations (i.e. groups of 8, 7, 6, 5, 4, 2 and single solutions). How many more dogs does he now need?

PROBABILITY

There are many ways to define (or fail to define!) probability and the reader with no knowledge of probability is advised to turn to a textbook. A probability is simply a number that we use to quantify the uncertainty surrounding an event. If we say an event has probability 0 then it is impossible, a probability of 0·5 (or $\frac{1}{2}$) means it is as likely to happen as not to (a 50–50 chance), and a probability of 1 means it is certain to happen.

One way of measuring probability is as follows:

Suppose an experiment can result in any one of n different outcomes, each equally likely to occur. If exactly s of these n outcomes constitute 'success' then the probability of 'success' is s/n and probability of 'failure' is $1 - (s/n)$.

Example

What is the probability of picking eight score-draws on the football pools if there are 10 in the 60 matches on the coupon? Number of ways of getting desired result (i.e. 8 from the 10) is

$$^{10}C_8 = \frac{10!}{2!8!}$$

Number of possible choices of 8 from 60 is

$$^{60}C_8 = \frac{60!}{52!8!}$$

$$\text{Probability} = \frac{10!/(2!8!)}{60!/(52!8!)} = \frac{10! \times 52!}{60! \times 2!}$$

Using the $\boxed{n!}$ key: 1.75×10^{-8}
Using Stirling's formula: 1.75×10^{-8} } i.e. 1 chance in 57 million

Some things to do

Exercise 8.4   Of 100 transistors in a batch, 13 are faulty.

(*a*) A technician checks a transistor at random, returns it to the batch, selects again, . . . and so on for 10 tests. What is the probability that no faulty transistor will be selected?

The formula for m tests from a batch of n where f are faulty is:

$$\text{probability of none detected} = \left(1 - \frac{f}{n}\right)^m$$

(*b*) A wiser technician does not replace the 10 transistors he tests. What is the probability of his not locating a faulty transistor?

$$\text{Formula}: \frac{^{n-f}C_m}{^nC_m}$$

(Hint: cancelling the factorials and multiplying may be quicker than using Stirling's formula.)

Exercise 8.5   The probability that k people all have different birthdays is

$$\frac{365!}{(365-k)!365^k}$$

The assumptions are that February 29th is ignored and all days in the year are equally likely birth dates.

By trying $k = 10, 20, 30 \ldots$ and then closer values, find how many people taken together would give probabilities of 0.5 and of 0.99 of finding common birthdays. (Use Stirling's formula for $n!$ or cancel down the factorials.)

Exercise 8.6 Years ago cigarette manufacturers included cards in their packets which could be collected to make sets (often 25 or 50 different cards to a set). More recently there have been cards with packets of tea, and with various confectionery.

(*a*) How many packets of chewing gum must one *expect* to buy (on average) to get a complete set of 25 different cards if each packet has one card?

The formula for this is, for n different cards in a set:

$$\text{Expected no.} = n\left(1 + \frac{1}{2} + \frac{1}{3} + \ldots + \frac{1}{n}\right)$$

(i) Evaluate this using the $\boxed{1/x}$ key.

(ii) Estimate the result using Euler's approximation for $1 + \frac{1}{2} + \ldots 1/n$ which is $\ln n + 0{\cdot}5772$.

(*b*) How many packets of tea would be needed for one to expect to get a complete set of 50 Teacards?

Exercise 8.7 (*a*) The Binomial expansion of $(1+x)^n$ for all n is

$$(1+x)^n = 1 + nx + \frac{n(n-1)}{2!}x^2 + \frac{n(n-1)(n-2)}{3!}x^3 + \ldots$$

For the case of n a positive integer, it terminates, and can be expressed as

$$(1+x)^n = {}^nC_0 + {}^nC_1 \times x + {}^nC_2 \times x^2 + \ldots + {}^nC_n \times x^n$$

The coefficients ${}^nC_0$, ${}^nC_1$, ${}^nC_2 \ldots {}^nC_n$ are the familiar Pascal's Triangle numbers.

Find the line of Pascal's Triangle for which $n = 10$.

(*b*) If x is small relative to 1 then the full expansion of $(1+x)^n$ is not needed for an accurate estimate of $(1+x)^n$ since the later terms are negligible.

For the case $(1+{\cdot}07)^{10}$ evaluate $1{\cdot}07^{10}$ directly using the $\boxed{y^x}$ key, and then use the Binomial expansion $1 + nx + (n(n-1)/2!)x^2 + \ldots$ adding successive terms until no change is noticed. Compare the results.

(*c*) Substituting $x = 1$ in the Binomial expansion for n a positive integer shows that

$$2^n = {}^nC_0 + {}^nC_1 + {}^nC_2 + \ldots {}^nC_n$$

Check this for various values of n.

BINOMIAL DISTRIBUTION

If the probability of some particular outcome ('success') to a 'trial' is p and there are n independent repetitions of the trial then the number of successes might be 0, or 1, or 2, or $\ldots$, or n.

For example, if at a maternity hospital there are 30 births in a week then the number of boys born could be any number from 0 to 30. Obviously the extremes are very unlikely because boys and girls are more or less equally likely ($p \approx \frac{1}{2}$). If however we were interested in the babies born with *spina bifida* ($p \approx 1/500$) then for there to be more than 0 or 1 out of the 30 would be very surprising. Clearly the probability of a certain number, r, occurring will depend upon r itself, upon the number of 'trials', n, and upon the individual probability, p.

The Binomial Probability Distribution is the pattern which such events tend to follow and the formula is

$$\text{probability of } r \text{ occurrences in } n \text{ trials} = \frac{n!}{r!(n-r)!} \times p^r(1-p)^{n-r}$$

Example 1

When a coloured rat and an albino rat are crossed the probability of the offspring of their offspring being coloured is $\frac{3}{4}$ and of being albino is $\frac{1}{4}$. What is the probability that a litter of seven will have three albinos? Here $n=7$, $r=3$, $p=\frac{1}{4}$.

$$\text{probability} = \frac{7!}{3!4!} \times (\cdot 25)^3(\cdot 75)^4 \approx 0 \cdot 173.$$

This is readily calculated whether $\boxed{n!}$ is available or not.
(*a*) Using $\boxed{n!}$:

Algebraic: 7 $\boxed{n!}$ $\boxed{\div}$ 3 $\boxed{n!}$ $\boxed{\div}$ 4 $\boxed{n!}$ $\boxed{=}$ $\boxed{\text{STO}}$ ·25 $\boxed{y^x}$
3 $\boxed{\times}$ ·75 $\boxed{x^2}$ $\boxed{x^2}$ $\boxed{\times}$ $\boxed{\text{RCL}}$ $\boxed{=}$

Reverse Polish: 7 $\boxed{n!}$ 3 $\boxed{n!}$ $\boxed{\div}$ 4 $\boxed{n!}$ $\boxed{\div}$ ·25 $\boxed{\text{enter}}$ 3
$\boxed{y^x}$ $\boxed{\times}$ ·75 $\boxed{x^2}$ $\boxed{x^2}$ $\boxed{\times}$

Since $\dfrac{7!}{3!4!} = \dfrac{7 \times 6 \times 5}{3 \times 2 \times 1} = 35$, $\boxed{n!}$ is not really necessary:

(*b*) Not using $\boxed{n!}$:

Algebraic: ·25 $\boxed{y^x}$ 3 $\boxed{\times}$ ·75 $\boxed{x^2}$ $\boxed{x^2}$ $\boxed{\times}$ 35 $\boxed{=}$

Reverse Polish: ·25 $\boxed{\text{enter}}$ 3 $\boxed{y^x}$ ·75 $\boxed{x^2}$ $\boxed{x^2}$ $\boxed{\times}$ 35 $\boxed{\times}$

Example 2

Approximately 26% of a country's population can taste diluted phenylthiocarbamide (PTC). A thousand random samples of 15 people each are taken at various places throughout the country. How many samples can be expected to contain 0, 1, ... 15 PTC tasters?

The probabilities of 0, 1, . . . 15 PTC tasters are given by the terms of

$$\frac{15!}{r!(15-r)!}(0\cdot26)^r(0\cdot74)^{15-r}$$

for $r=0, 1 \ldots 15$.

These can be readily evaluated using a calculator and the results are as shown in Table 8.1, accurate to 3 d.p. Probabilities for $r=11$ to 15 are negligible.

TABLE 8.1

r	0	1	2	3	4	5	6	7	8	9	10
$p(r)$	0·011	0·058	0·142	0·216	0·227	0·176	0·103	0·046	0·016	0·004	0·001

The method of computing these values should be noted. It is not necessary to perform the calculation for each r value separately although this is not particularly arduous.

The terms needed are:

r	0	1
$p(r)$	$(0\cdot74)^{15}$	$\dfrac{15}{1}\times(0\cdot26)\times(0\cdot74)^{14}$

r	2	3
$p(r)$	$\dfrac{15\times14}{1\times2}\times(0\cdot26)^2\times(0\cdot74)^{13}$	$\dfrac{15\times14\times13}{1\times2\times3}\times(0\cdot26)^3\times(0\cdot74)^{12}$

and so on.

Thus to get from one term to the next requires only two multiplications and two divisions at most.

Procedure

(i) $(\cdot74)^{15}$ is calculated.

(ii) Multiply by 15 and by $\cdot26$; Divide by 1 and by $\cdot74$

(iii) Multiply by 14 and by $\cdot26$; Divide by 2 and by $\cdot74$

$\vdots$ $\qquad\qquad$ $\vdots$ $\qquad\qquad$ $\vdots$

(xvi) Multiply by 1 and by $\cdot26$; Divide by 15 and by $\cdot74$

(Obviously this can be slightly streamlined in practice.)

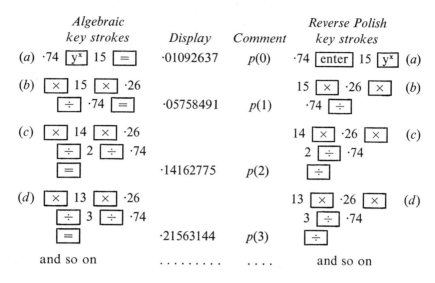

	Algebraic key strokes	*Display*	*Comment*	*Reverse Polish key strokes*	
(a)	·74 $\boxed{y^x}$ 15 $\boxed{=}$	·01092637	$p(0)$	·74 $\boxed{enter}$ 15 $\boxed{y^x}$	(a)
(b)	$\boxed{\times}$ 15 $\boxed{\times}$ ·26 $\boxed{\div}$ ·74 $\boxed{=}$	·05758491	$p(1)$	15 $\boxed{\times}$ ·26 $\boxed{\times}$ ·74 $\boxed{\div}$	(b)
(c)	$\boxed{\times}$ 14 $\boxed{\times}$ ·26 $\boxed{\div}$ 2 $\boxed{\div}$ ·74 $\boxed{=}$	·14162775	$p(2)$	14 $\boxed{\times}$ ·26 $\boxed{\times}$ 2 $\boxed{\div}$ ·74 $\boxed{\div}$	(c)
(d)	$\boxed{\times}$ 13 $\boxed{\times}$ ·26 $\boxed{\div}$ 3 $\boxed{\div}$ ·74 $\boxed{=}$	·21563144	$p(3)$	13 $\boxed{\times}$ ·26 $\boxed{\times}$ 3 $\boxed{\div}$ ·74 $\boxed{\div}$	(d)
	and so on	. . . .	. . . .	and so on	

Actually a further economy can be made by first evaluating ·26/·74 and storing the value and recalling it each time. For example line (c) would become

(c) (*Algebraic*) $\boxed{\times}$ 14 $\boxed{\times}$ RCL $\boxed{\div}$ 2 $\boxed{=}$

or (*Reverse Polish*) 14 $\boxed{\times}$ RCL $\boxed{\times}$ 2 $\boxed{\div}$

Multiplying the values for the probabilities by 1000 gives the expected distribution (fig. 8.1) which is quite symmetrical despite the fact that $p\ (=0·26)$ is well away from the 'symmetrical' value 0·5.

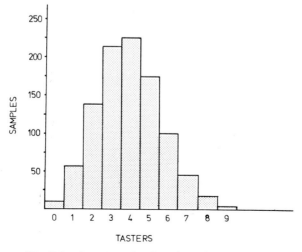

Fig. 8.1   An example of a Binomial Distribution.

The resemblance between the Binomial Distribution for large n and the Normal Curve is important. The Normal Distribution is the limiting case of the Binomial Distribution, as n increases without end.

Some things to do

Exercise 8.8   The most likely number of heads to occur when a coin is tossed 50 times is 25. How likely is that to happen?

Exercise 8.9   The incidence of *spina bifida* in newly born children is about 1 in 500. What is the probability that of the first 800 babies born in a new hospital the number with *spina bifida* will be (*a*) 0; (*b*) 1; (*c*) 2; (*d*) more than 2?

Exercise 8.10   An electronic digital computer has 128000 ferrite memory cores. The probability of failure of any one of these cores on any day is 1 in a million. What is the probability that there will be no failure in 10 days running?

Exercise 8.11   An injection to protect pregnant women from *rubella* is known to be successful in 98% of cases where the woman is exposed to the disease. Complete a table of probabilities for $r = 0, 1, 2, 3, 4, 5, 6$ women still catching *rubella* out of 100 treated women who are exposed to the disease. Plot a graph of this skewed (asymmetrical) distribution.

POISSON DISTRIBUTION

The Binomial Distribution requires a knowledge of the probability that some event will happen (p). The probability that it will *not* happen $(1-p)$ is of course also then known and is needed in the formula. The probability p is often derived from finding the proportion of 'successes' in n 'trials'. Now there are many situations where we can count the occurrences of some event but no figure for non-occurrences is meaningful (we cannot attach any meaning to the idea of the proportion of successes). For example, a geiger counter records 7 counts in 20 seconds. How many counts do *not* occur? Obviously we do not know. As another example a botanist throws a 1 metre diameter hoop 'at random' in a field and counts 3 buttercup plants growing there. How many buttercup plants are *not* growing there? Again no answer is possible. Obviously the Binomial Distribution is inapplicable to such problems.

In such situations the Poisson Distribution may be useful. It derives from the series expansion of the exponential function e^x.

The probabilities for 0, 1, 2, . . . occurrences are given by successive terms of

$$e^{-x}\left(1 + x + \frac{x^2}{2!} + \frac{x^3}{3!} + \frac{x^4}{4!} \cdots\right)$$

where x is the average (or expected) number of occurrences of the event of interest. In other words the probability of no occurrences is e^{-x}, of 1 occurrence is xe^{-x}, of 2 occurrences is $(x^2/2!)e^{-x}$, of 3 occurrences is $(x^3/3!)e^{-x}$, and so on.

Example 1

We now look at a case where only the average number of occurrences is known and the Binomial model cannot be applied because no probabilities can be assigned.

Breakdowns of an electronic digital computer average out at 2·3 per week and appear to occur randomly. Find the probabilities that in a given week there will be 0, 1, 2, 3, 4, more than four, breakdowns.

Here $x = 2\cdot3$. The Poisson model gives the probability of r breakdowns as $p(r) = (x^r/r!)e^{-x}$. (see page 154 for details of the calculations.)

Example 2

We will now look again at the *rubella* problem 8.11 which was to be solved using the Binomial Distribution.

The probability of a treated woman catching *rubella* if exposed to it is 0·02. If we study the cases of 100 such women, how many *rubella* cases are likely to occur? Obviously the *expected* (or average) number will be $100 \times 0\cdot02 = 2$ cases. This is our x.

Using Poisson we see that the probabilities $(p(r) = x^r e^{-x}/r!)$ of r cases will be (to 3 d.p.).

$$r = 0 \qquad p(0) = e^{-2} = 0\cdot135$$

$$r = 1 \qquad p(1) = 2e^{-2} = 0\cdot271$$

$$r = 2 \qquad p(2) = \frac{2^2}{2!} e^{-2} = 0\cdot271$$

$$r = 3 \qquad p(3) = \frac{2^3}{3!} e^{-2} = 0\cdot180$$

and so on. These values are simply computed using the $\boxed{e^x}$ key of a calculator. Note that the next value $p(r+1)$ can be found from the preceding value $p(r)$ by multiplying by $x (=2)$ and dividing by r. The reader is invited to devise a suitable procedure.

The results obtained by Poisson and by Binomial are compared in Table 8.2.

TABLE 8.2

	0	1	2	3	4	5	6	7
Poisson	·135	·271	·271	·180	·099	·036	·012	·003
Binomial	·133	·271	·273	·182	·090	·035	·011	·003

Key strokes

	Results to 3 d.p.	*Algebraic*	*Reverse Polish*
$p(0) = e^{-2\cdot3}$	$= 0\cdot100$	2·3 [+/−] [eˣ] [STO]	2·3 [+/−] [eˣ] [STO]
$p(1) = 2\cdot3\,e^{-2\cdot3}$	$= 0\cdot231$	[×] 2·3 [=] [M+]	2·3 [×] [Σ]
$p(2) = \dfrac{(2\cdot3)^2}{2!}\,e^{-2\cdot3}$	$= 0\cdot265$	[×] 2·3 [÷] 2 [=] [M+]	2·3 [×] 2 [÷] [Σ]
$p(3) = \dfrac{(2\cdot3)^3}{3!}\,e^{-2\cdot3}$	$= 0\cdot203$	[×] 2·3 [÷] 3 [=] [M+]	2·3 [×] 3 [÷] [Σ]
$p(4) = \dfrac{(2\cdot3)^4}{4!}\,e^{-2\cdot3}$	$= 0\cdot117$	[×] 2·3 [÷] 4 [=] [M+]	2·3 [×] 4 [÷] [Σ]
These add up to	$\overline{0\cdot916}$ so		[[last x] may be needed after each [Σ]]
$p(5+) = 1 - 0\cdot916$	$= 0\cdot084$	1 [−] [RCL] [=]	1 [RCL] [−]

The required probabilities, then, are 0·100, 0·231, 0·265, 0·203, 0·117, 0·084.

It is clear from Table 8.2 that in cases where the individual probability is small but the number of cases large, the Poisson Distribution is a very good approximation to the Binomial Distribution—giving a characteristically skewed (asymmetrical) distribution (fig. 8.2).

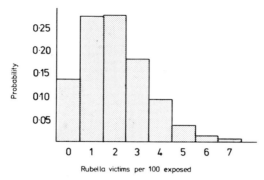

Fig. 8.2   An example of a Poisson Distribution.

Something to do

Exercise 8.12   The incidence of *spina bifida* was the subject of Exercise 8.9. For 800 births the expected number of such cases is $1/500 \times 800 = 1\cdot6$. Use the Poisson distribution to estimate the probabilities of (*a*) 0; (*b*) 1; (*c*) 2; (*d*) more than 2 cases, in the first 800 births at a hospital. Compare your results with those obtained using the Binomial distribution.

9. Investigating graphs and relationships

WHAT IS A FUNCTION?

Very simply, a function is a rule (or equation) which relates one set of values to another.

A more precise definition is: a function is a rule which associates with any given number in some set (the 'domain') one number from some other set (the 'codomain'). As an example we take the function, f, 'square and then add 3' with domain the set of all real numbers.

We can write this as

$$f(x) = x^2 + 3$$

Clearly the pocket calculator is often useful for evaluating particular function values: e.g. for the above function

$$f(3 \cdot 72) = 3 \cdot 72^2 + 3 = 16 \cdot 8384$$

(In this example the possible range of $f(x)$ values—the codomain—is the set of all real numbers $\geqslant +3$.)

PLOTTING GRAPHS

Graphs often have the y-axis representing the function values $f(x)$, corresponding to x values. Given the task of plotting the curve

$$y = f(x) = x^2 e^{-x} + \sin x \quad \text{for } 0 \leqslant x \leqslant 5$$

the calculator can be used to quickly evaluate $f(x)$ at suitable x values (perhaps every $0 \cdot 5$) and so provide a table of values from which a sketch can be drawn. The calculator is so efficient at this kind of computation that what would otherwise be a considerable chore becomes almost trivial. (A table for $y = x^2 e^{-x} + \sin x$ will be found in 'Solutions and Notes'.)

Behaviour of Functions

It is of interest to plot the way in which functions such as $x \sin x$, $x^n e^{-x}$, $\sin (1/x)$, $(\ln x)/x$ behave when x varies.

Example 1

Taking $f(x) = x^2 e^{-x}$, as x gets large x^2 will get large but e^{-x} will get small. Which will 'win'? I.e. does the function tend to a *large* value or to a *small* value as x increases? Using a calculator the behaviour is soon revealed (Table 9.1 and fig. 9.1).

TABLE 9.1

x	1	2	3	4	5	6
$f(x)$	·368	·541	·448	·293	·168	·089

x	7	8	9	10	100
$f(x)$	·045	·021	·001	·0005	$3\cdot7 \times 10^{-40}$

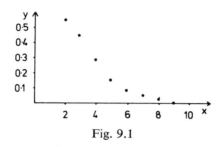

Fig. 9.1

Example 2

Taking $f(x) = (x+1)/(x-2)$ how does the function behave as x goes from 0 to $+\infty$? We would need this information in order to sketch the graph of $y = (x+1)/(x-2)$. We can calculate $f(x)$ for particular values of x:

TABLE 9.2

x	0	1	2	3	4	5
$f(x)$	$-0\cdot500$	$-2\cdot000$	?	$4\cdot000$	$2\cdot500$	$2\cdot000$

x	10	50	100	500	1000	10000
$f(x)$	$1\cdot375$	$1\cdot063$	$1\cdot031$	$1\cdot006$	$1\cdot003$	$1\cdot000$

The function is clearly tending to a value of 1 as $x \to +\infty$ (i.e. as x gets indefinitely large). However the value at $x=2$ cannot be found since it implies division by zero (3/0) which is not defined. Therefore the function as given by the expression $(x+1)/(x-2)$ is not defined at $x=2$. At any such point where the denominator is zero some non-direct method is needed to investigate the behaviour.

LIMITS

What we *can* do to study the behaviour of $(x+1)/(x-2)$ at $x=2$ is to take values *near to* $x=2$ and see what happens as we get closer and closer to 2, i.e. as 'x tends to 2' ($x \to 2$).

The sequence of estimates in Table 9.3 shows that as x tends to 2 from the right ($x \to 2^+$) then $(x+1)/(x-2)$ gets large, i.e.

$$\frac{x+1}{x-2} \to +\infty \quad \text{as } x \to 2^+$$

TABLE 9.3

x	2·5	2·1	2·01	2·001	2·0001	2·00001
$f(x)$	7	31	301	3001	30001	300001

Do we get the same result if we approach 2 from the left? The sequence of estimates in Table 9.4

TABLE 9.4

x	+1·5	1·9	1·99	1·999	1·9999
$f(x)$	−5	−29	−299	−2999	−29999

shows that as $x \to 2^-$ (x tends to 2 from the left) $(x-1)/(x-2) \to -\infty$ (i.e. gets indefinitely large negatively). These results are illustrated in fig. 9.2.

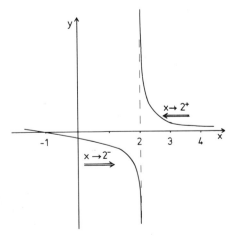

Fig. 9.2

This behaviour can be expressed in terms of *limits*:

$$\lim_{x \to 2^+} \left(\frac{x+1}{x-2}\right) = +\infty$$

$$\lim_{x \to 2^-} \left(\frac{x+1}{x-2}\right) = -\infty$$

We shall return to look further at limits later in the chapter.

Some things to do

Exercise 9.1   Investigate the behaviour of the following functions, taking all real numbers as the domain, where possible, and so sketch their graphs:

(a) xe^{-x} (b) $x \sin(x)$ (c) $\dfrac{2x+3}{x+4}$ (d) $\dfrac{x^2}{x-1}$ (e) e^{-x^2}

Exercise 9.2   According to Einstein's Special Theory of Relativity the mass m of a body increases with its velocity, v, such that

$$m = m_0/\sqrt{(1-v^2/c^2)}$$

where m_0 is the mass of the body at rest and c is the speed of light in vacuum ($c \approx 3 \times 10^8$ m s^{-1}). For the case $m_0 = 1$ kg plot the graph of m against v (taking $v=0$ to $v=c$ in steps of $c/10$), to show how the mass varies with the velocity.

Exercise 9.3   The power dissipated in the resistor shown in the diagram is given by: power $= E^2 R/(R+r)^2$ where r is the internal resistance of the battery. By taking the case when $r=6$ Ω, $E=1\cdot5$ V, plot power against R (for $R=0$ to 20 in steps of 1) and find when the power dissipated is a maximum. What do you notice? Is the result generally true?

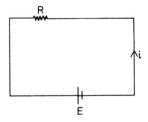

Exercise 9.4   Physiologists have found that during coughing the air-passage diameter is decreased. Theoretical reasoning (due to B. F. Visser) suggests that the velocity, v, of the expelled air through a tube of normal radius r_0 but contracted to a radius of r is given by

$$v = k(r^2 r_0 - r^3)$$

It may be assumed that the contraction of the air passage is to increase the velocity of the expelled air. By plotting the graph of v against r, taking $k=1$ and $r_0=1$ find the radius r to maximize the velocity. Can you interpret your findings?

Exercise 9.5   When a subject is engaged in strenuous work oxygen may be taken from the muscles faster than it is being replaced (giving an 'oxygen debt'). When the work ceases the oxygen debt is slowly reduced. A suggested mathematical model of this, for a constant work rate, is:

(i) *Work phase*, $t=0$ to $t=t_1$

$$y=k(1-e^{-t/\tau})$$

(ii) *Recovery phase*, $t=t_1$ to $t=t_2$

$$y=y_1 e^{-(t-t_1)/\tau}$$

where y is the rate of intake of extra oxygen
   k is some constant
   t is the time, starting from when the work begins
   τ is a time constant
   y_1 is the value of y at time $t=t_1$ (i.e. when the work ceases).

(a) Plot the curve for (i) taking $\tau=4$, $t_1=10$, $k=1$.

(b) Using the value of y_1 now known, plot the curve for (ii) on the same graph, for $t_1=10$, to $t_2=20$.

Exercise 9.6   The Maxwell distribution of molecular speeds in a gas is:

$$f(v)=4\pi(a/\pi)^{3/2}v^2 e^{-av^2}$$

where $a=M/2kT$ (kg J^{-1})
   M is the molecular mass (kg)
   k is the Boltzmann constant ($1 \cdot 38 \times 10^{-23}$ JK^{-1})
   T is the temperature (K)
   v is the velocity of a molecule (ms^{-1}).

(a) For a particular choice of T and M we get a velocity distribution curve shaped rather like a bell. For the case of hydrogen molecules ($M=1 \cdot 67 \times 10^{-27}$ kg) plot the velocity distribution curve for temperature $T=400$ K, i.e. plot $y=4\pi(a/\pi)^{3/2}v^2 e^{-av^2}$ for $v=0$ to $v=10000$ (say).

(b) If the temperature is increased what effect will it have on the velocities? Using the same axes plot the curve for $T=600$ K to check your reasoning.

Exercise 9.7   In a modified form of Searle's apparatus, described by Lorrimer, McMullan and Walmley (1976) a very thin cylindrical metal rod, surrounded by a cylindrical insulating jacket, is electrically heated at one end and held in a water-cooled copper block at the other end. The temperatures at the two ends of the rod are measured by means of thermocouples. Theory shows that the temperature, T_x, at any point along the rod is given by

$$T_x = T_r + (T_0 - T_r)e^{-gx} + [T_l - T_0 e^{-gl} + T_r(1 - e^{-gl})]x/l$$

where T_r is the room temperature

T_0, T_l are the rod temperatures at the hot end $(x=0)$ and cold end $(x=l)$

$$g^2 = \frac{k_2}{k_1} \frac{2}{a^2 \ln (r/a)}$$

a is the radius of the rod
r is the outer radius of the insulating jacket
l is the rod length
k_1, k_2 are the thermal conductivities of rod and insulation.

The formula for T_x is accurate provided k_2/k_1 is much smaller than 1.
(a) Plot the temperature gradient along the rod for the following two cases where $T_r = 25°C$, $T_0 = 5°C$, $T_l = 100°C$, $l = 200$ mm.

(i) $k_2/k_1 = 10^{-4}$, pure metal rod with good insulation material
$a = 5$ mm
$r = 60$ mm
(ii) $k_2/k_1 = 10^{-3}$, alloy rod with good insulation material
$a = 0.5$ mm
$r = 60$ mm

(b) Try some other values for a and r and interpret the results.
Exercise 9.8 (a) The axial field of a plane circular coil at a distance x from the centre of the coil is given by

$$B(x) = \frac{\mu_0 n I r^2}{2(r^2 + x^2)^{3/2}}$$

For the typical values $\mu_0 = 1.257 \times 10^{-6}$ Hm^{-1}
$n = 320$
$I = 1.0$ A
$r = 0.07$ m

complete the table:

x	0	0·01	0·02	0·03	0·04	0·05	0·06	0·07
$B(x)$								

(First calculate $\mu_0 n I r^2/2$ and store it in a memory or in the stack.)

(b) Often two identical coils are placed side by side (parallel) in order to produce a fairly uniform field along the axis between the coils the distance $r=0.07$ m being the normal separation distance used. Such an arrangement is known as a Helmholtz coil system.

The total field strength is found by adding the two separate contributions at each point x. This is very easily achieved using the above table. All that is required is to write down the strengths $B(x)$ in reverse order (i.e. result of $B(0.07)$ under $B(0)$, $B(0.06)$ under $B(0.01)$ and so on) and then add. Do this and plot the total field strength against x to see how uniform it is. Investigate other separation distances.

LIMIT APPROACHING A POINT

We have seen, in the case of $y=f(x)=(x+1)/(x-2)$ that as we approach 2 from the left the value of $(x+1)/(x-2)$ tends towards $-\infty$, and approaching from the right it tends towards $+\infty$. There is a left limit and a right limit but they differ.

For a limit to exist as $x\to a$ we need three conditions to hold:

(i) $\lim_{x\to a^-}[f(x)]=L$ (some number, the left limit, must exist)

(ii) $\lim_{x\to a^+}[f(x)]=R$ (some number, the right limit, must exist)

(iii) $L=R$ (the two limits must be equal)

i.e. both left and right limits must exist and they must be the same. If we have these three conditions holding then we write

$$\lim_{x\to a}[f(x)]= \ldots$$

Example 1

For the function $g(x)=(x^3+x-2)/(x-1)$ we use the pocket calculator to investigate $\lim_{x\to1^+}[g(x)]$ and $\lim_{x\to1^-}[g(x)]$. Tables 9.5A and 9.5B can be drawn up:

TABLE 9.5A $x\to1^-$

x	0·9	0·99	0·999	0·9999	0·99999	0·999999
$g(x)$	3·71	3·9701	3·997001	3·9997	3·99997	3·999997

TABLE 9.5B $x\to1^+$

x	1·1	1·01	1·001	1·0001	1·00001	1·000001
$g(x)$	4·31	4·0301	4·003001	4·0003	4·00003	4·000003

From these results we conclude that

$$\text{Lim}_{x \to 1^-} [g(x)] = 4$$

$$\text{Lim}_{x \to 1^+} [g(x)] = 4$$

Since these limits both exist and are the same, we can say that there is a limit as $x \to 1$ and we write: $\text{Lim}_{x \to 1} [g(x)] = 4$.

Now the function $g(x) = (x^3 + x - 2)/(x - 1)$ is undefined at $x = 1$. We cannot put $x = 1$ in to find $g(1)$ because that would make the denominator zero. We could arbitrarily define $g(1)$ to be anything we like. A sensible choice would be $g(1) = 4$ but if we wished we could define $g(1) = 0 \cdot 99$ for example.

CONTINUITY

For a graph to be continuous at a point it must not suddenly jump leaving a gap. At any point we must have the limit existing (conditions (i), (ii), (iii) must hold) but also we need

(iv) $f(a) = \text{Lim}_{x \to a} [f(x)]$.

In other words the function must be defined at the point, $x = a$, and its value must equal that of the limit as $x \to a$.

So for the function defined as

$$g(x) = \frac{x^3 + x - 2}{x - 1} \quad \text{provided } x \neq 1$$
$$= 4 \quad \quad \text{if } x = 1$$

we can say the *limit exists as* $x \to 1$ and since $\text{Lim}_{x \to 1} [g(x)] = 4 = g(1)$ the function *is continuous at* $x = 1$ (fig. 9.3).

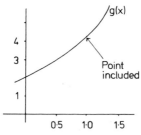

Fig. 9.3   Continuous.

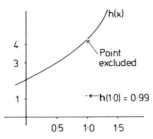

Fig. 9.4   Not continuous.

However, if we have another function defined by

$$h(x) = \frac{x^3 + x - 2}{x - 1} \quad \text{provided } x \neq 1$$
$$= 0 \cdot 99 \qquad \text{if } x = 1$$

then the function value (0·99) and the limit value (4), for $x = 1$, are not the same, so the function *is not continuous at* $x = 1$ (fig. 9.4).

Example 2

If we take $h(x) = \sin(1/x)$ what happens as $x \to 0$? Tables 9.6A and 9.6B show the situation:

TABLE 9.6A $x \to 0^-$

x	$-\cdot1$	$-\cdot01$	$-\cdot001$	$-\cdot0001$	$-\cdot00001$	$-\cdot000001$
$h(x)$	$+\cdot544$	$+\cdot506$	$-\cdot827$	$+\cdot306$	$-\cdot036$	$+\cdot350$

TABLE 9.6B $x \to 0^+$

x	$\cdot1$	$\cdot01$	$\cdot001$	$\cdot0001$	$\cdot00001$	$\cdot000001$
$h(x)$	$-\cdot544$	$-\cdot506$	$+\cdot827$	$-\cdot306$	$+\cdot036$	$-\cdot350$

There is no obvious limit from either side and so no limit exists as $x \to 0$. Whatever we define $h(0)$ to be we cannot have a limit at $x = 0$ and so the function cannot be continuous at $x = 0$.

Some things to do

Exercise 9.9   Investigate the existence of limits and continuity of the following functions at the points indicated. (In each case the real numbers form the domain).

Function Definition		Point
(a) $f(x)=\dfrac{x^2+x-12}{x-3}$ $=7$	provided $x\neq 3$ when $x=3$	$x=3$
(b) $f(x)=x+\sin x$	for all x	$x=-2$
(c) $f(x)=x^6-x^4+x^3$	for all x	$x=2$
(d) $f(x)=\dfrac{x^2+5x}{1-x^3}$ $=100$	if $x\neq 1$ if $x=1$	$x=5$ and $x=1$
(e) $f(x)=\dfrac{\sin x}{x}$ $=1$	if $x\neq 0$ if $x=0$	$x=0$
(f) $f(x)=\dfrac{\tan x}{x}$ $=0$	if $x\neq 0$ if $x=0$	$x=0$
(g) $f(x)=\cos(1/x)$ $=1$	if $x\neq 0$ if $x=0$	$x=0$
(h) $f(x)=\dfrac{x^4}{e^x-1}$ $=0$	if $x\neq 0$ if $x=0$	$x=0$
(i) $f(x)=e^{-(1/x^2)}$ $=1$	if $x\neq 0$ if $x=0$	$x=0$
(j) $f(x)=\dfrac{\ln(1+x^2)}{x^2}$ $=1$	if $x\neq 0$ if $x=0$	$x=0$
(k) $f(x)=\dfrac{x^4e^x}{(e^x-1)^2}$ $=0$ (This function occurs in Exercise 13.15.)	if $x\neq 0$ if $x=0$	$x=0$
(l) $f(x)=\arctan(1/x)$ $=0$	if $x\neq 0$ if $x=0$	$x=0$

Exercise 9.10   The perimeter of an *n*-sided regular polygon inscribed in a circle of radius $\frac{1}{2}$ is

$$P = n \sin \left(\frac{180°}{n} \right)$$

As *n* increases the polygon and the circle become indistinguishable. Show, by taking large values for *n*, that

$$\operatorname*{Lim}_{n \to \infty} [n \sin (180°/n)] = \pi$$

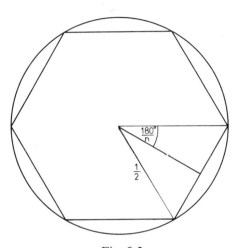

Fig. 9.5

10. Solving non-linear equations

INTRODUCTION

Much scientific work involves the setting up of a mathematical model. In other words, a scientific problem is turned into a mathematical problem which hopefully takes account of all the important scientific features. The mathematical problem is then solved and the results interpreted in terms of the original scientific problem. This sometimes leads to a non-linear equation, examples of which are $x^5 + 10x^2 - 99 = 0$; $x = \tan x$; and $\sin x = e^x$. A non-linear equation $f(x) = 0$ is one for which the graph $y = f(x)$ is not a straight line. Solving such an equation can be fairly straightforward if a formula exists but this is rarely the case and approximate numerical techniques are usually needed. This chapter begins by considering some of the limited formulae available and then describes some of the simpler numerical methods which are often employed.

FORMULAE FOR SOLVING NON-LINEAR EQUATIONS

QUADRATIC EQUATIONS

To find the roots of the quadratic equation $ax^2 + bx + c = 0$, the quadratic formula

$$x = \frac{-b \pm \sqrt{(b^2 - 4ac)}}{2a}$$

may be used.

Example 1

Find the two roots x_1, x_2 of $3{\cdot}71x^2 + 12{\cdot}9x + 2{\cdot}76 = 0$

$$x_1 = \frac{-12{\cdot}9 + \sqrt{(12{\cdot}9^2 - 4 \times 3{\cdot}71 \times 2{\cdot}76)}}{2 \times 3{\cdot}71}$$

$$x_2 = \frac{-12{\cdot}9 - \sqrt{(12{\cdot}9^2 - 4 \times 3{\cdot}71 \times 2{\cdot}76)}}{2 \times 3{\cdot}71}$$

(It is usually best when using a calculator to evaluate the $\sqrt{}$ part first and for algebraic machines that is most simply done as $(-4 \times 3 \cdot 71 \times 2 \cdot 76 + 12 \cdot 9^2)^{\frac{1}{2}}$, i.e. $\sqrt{(-4ac + b^2)}$

$$\therefore \quad x_1 \approx -0 \cdot 22904072 \qquad x_2 \approx -3 \cdot 2480482$$

In calculating x_2 apparently the whole calculation must be repeated with a minus sign replacing the plus sign. However, various short cuts are possible. One trick is to store the result of the $\sqrt{(b^2 - 4ac)}$ calculation when finding x_1 and simply call it back when required again for x_2. Another method relies on the fact that $x_1 + x_2 = -b/a$, so $x_2 = (-b/a) - x_1 \approx -3 \cdot 2480482$ in this example. Thus the amount of computation can be considerably reduced. A third method uses the fact that $x_1 x_2 = c/a$, $\therefore x_2 = c/a x_1 \approx -3 \cdot 2480482$. Again quite a saving of effort is made.

It is possible to buy calculators which solve quadratic equations automatically (e.g. Commodore M55) and programmable calculators can easily be programmed to do this.

Example 2

Solve $1 \cdot 44x^2 - 7 \cdot 992x + 11 \cdot 0889 = 0$.

Using the formula we find $b^2 - 4ac$ is 0, which indicates two equal roots ('repeated roots'), so

$$x_1 = x_2 = 2 \cdot 775 \text{ (exactly)}.$$

Example 3

Solve $1 \cdot 44x^2 - 7 \cdot 900 + 11 \cdot 0889 = 0$.

Using the formula we find $b^2 - 4ac$ is $-1 \cdot 462064$ and since this is negative we cannot find its square root. (To try produces an error message on the calculator.) This indicates that the roots are complex numbers—no real roots exist.

The roots in this case are

$$x_1 = \frac{-b}{2a} + \frac{\sqrt{(4ac - b^2)}}{2a} j, \quad x_2 = \frac{-b}{2a} - \frac{\sqrt{(4ac - b^2)}}{2a} j$$

where j is the square root of minus one $(\sqrt{-1})$.

Some things to do

Exercise 10.1   Solve the following quadratic equations by all the methods suggested and compare results for both accuracy and time taken and key strokes needed.

(a) $3 \cdot 792x^2 - 4 \cdot 279x + 0 \cdot 821 = 0$
(b) $x^2 - 7927 \cdot 0002x + 1 \cdot 5854 = 0$

Exercise 10.2   Calculate the percentage ionization of a 0·0011 M solution of formic acid, at 25°C, for which the equilibrium equation is

$$HCO_2H \rightleftharpoons H^+ + HCO_2{}^-.$$

Method

The equation is

$$\frac{[H^+][HCO_2{}^-]}{[HCO_2H]} = K_a$$

where $K_a = 1·6 \times 10^{-4}$ mol l^{-1}, and the symbol [ ] means the 'concentration of ions'—so [H$^+$] means the concentration of hydrogen ions, for example.

If x is the concentration of hydrogen ions, then

$$[H^+] = x, \quad [HCO_2{}^-] = x, \quad [HCO_2H] = 0·0011 - x$$

∴ we must solve

$$\frac{x^2}{0·0011 - x} = 1·6 \times 10^{-4}$$

This gives the quadratic equation $x^2 + 1·6 \times 10^{-4}x - 1·76 \times 10^{-7} = 0$. Solve this to find x and so find the % concentration of hydrogen ions which is $(x/0·0011) \times 100$.

CUBIC EQUATIONS

It is unfortunate that there is no simple formula for solving cubic equations. Although formulae do exist they are rather difficult to use and require a good deal of algebra and insight.

It can be shown that the general cubic equation $aX^3 + bX^2 + cX + d = 0$ can always be transformed (by the substitution $x = X + b/3a$) to the simpler form $x^3 + qx + r = 0$, and this is the starting point for the methods described here.

Cardan's solution of $x^3 + qx + r = 0$, for one real root

Provided that there is just one real root, the solution is given by

$$x = \left\{ \frac{-r}{2} + \sqrt{\left(\frac{r^2}{4} + \frac{q^3}{27} \right)} \right\}^{\frac{1}{3}} + \left\{ \frac{-r}{2} - \sqrt{\left(\frac{r^2}{4} + \frac{q^3}{27} \right)} \right\}^{\frac{1}{3}}$$

The calculator can be used with great effect here, the precise procedure followed depending on the machine used. It is generally best to evaluate the $\sqrt{}$ part first. If $(r^2/4) + (q^3/27) < 0$ indicating *three* real roots, Cardan's method fails and the method described later should be used.

Example

Solve Wallis's equation: $x^3 - 2x - 5 = 0$.
  Here $q = -2$, $r = -5$.

$$x = \left\{ \frac{5}{2} + \sqrt{\left(\frac{25}{4} - \frac{8}{27} \right)} \right\}^{\frac{1}{3}} + \left\{ \frac{5}{2} - \sqrt{\left(\frac{25}{4} - \frac{8}{27} \right)} \right\}^{\frac{1}{3}}$$

$$\approx \{2 \cdot 5 + 2 \cdot 4400213\}^{\frac{1}{3}} + \{2 \cdot 5 - 2 \cdot 4400213\}^{\frac{1}{3}}$$

$$\approx 2 \cdot 0945515.$$

Trigonometrical solution of $x^3 + qx + r = 0$, for three real roots

If the expression $(r^2/4) + (q^3/27)$ is negative then its square root is an imaginary number and Cardan's method is unsuitable. The three roots are written as:

$$x = \{a + jb\}^{\frac{1}{3}} + \{a - jb\}^{\frac{1}{3}}$$

where

$$a = -r/2$$

and

$$b = \sqrt{-\left(\frac{r^2}{4} + \frac{q^3}{27} \right)}$$

In this case it is necessary to find an angle θ such that $\tan \theta = b/a$, i.e. $\theta = \arctan (b/a)$, then the three solutions are:

$$\pm 2(a^2 + b^2)^{\frac{1}{6}} \cos \left(\frac{\theta}{3} \right); \quad \pm 2(a^2 + b^2)^{\frac{1}{6}} \cos \left(\frac{\theta + 2\pi}{3} \right);$$

$$\pm 2(a^2 + b^2)^{\frac{1}{6}} \cos \left(\frac{\theta + 4\pi}{3} \right)$$

(plus signs taken if a is positive, minus signs taken if a is negative).

Example

Find the roots of $x^3 = 15x + 4$.
  Rearranging: $x^3 - 15x - 4 = 0$.
  So $q = -15$, $r = -4$.

$$x = \left\{ \frac{4}{2} + \sqrt{\left(\frac{16}{4} - \frac{3375}{27} \right)} \right\}^{\frac{1}{3}} + \left\{ \frac{4}{2} - \sqrt{\left(\frac{16}{4} - \frac{3375}{27} \right)} \right\}^{\frac{1}{3}}$$

$$x = \{2 + \sqrt{-121}\}^{\frac{1}{3}} + \{2 - \sqrt{-121}\}^{\frac{1}{3}}$$

As the square roots here are imaginary we must use the trigonometric method.

Here, $a = 2$ (positive), $b = 11$, $a^2 + b^2 = 125$

$$\theta = \arctan(11/2) \approx 1.3909428$$

$$+ 2(a^2 + b^2)^{\frac{1}{2}} \cos\left(\frac{\theta}{3}\right) \approx 2 \times 2.2360680 \times \cos\left(\frac{1.3909428}{3}\right)$$

$$\approx +4.0000000$$

$$+ 2(a^2 + b^2)^{\frac{1}{2}} \cos\left(\frac{\theta + 2\pi}{3}\right) \approx -3.7320508$$

$$+ 2(a^2 + b^2)^{\frac{1}{2}} \cos\left(\frac{\theta + 4\pi}{3}\right) \approx -0.26794919$$

This provides a good example of where the arctan function is needed in addition to the ordinary trigonometric functions.

The awkwardness of the above procedures for those without calculators has led to their being neglected in the past and now iterative procedures are emphasized much more.

A rather different approach to solving these equations on a calculator can be found in J. M. Smith (1975).

QUARTIC AND HIGHER DEGREE EQUATIONS

Generally quartic equations ($ax^4 + bx^3 + cx^2 + dx + e = 0$) are even more awkward to solve analytically (i.e. by direct solution by formula) and it has been shown that no general formulae exist at all for higher degree equations. Thus numerical methods, as we next describe, are important for solving higher degree polynomial equations as well as transcendental equations (involving trigonometric, hyperbolic, exponential or logarithmic functions).

Some things to do

Exercise 10.3   The parabola $y = 2x^2 + 5x - 3$ and the cubic curve $y = x^3 + 2x^2 - x - 2$ intersect at one, two or three points. By solving the cubic equation

$$2x^2 + 5x - 3 = x^3 + 2x^2 - x - 2$$

find the point(s) of intersection.

Exercise 10.4   Find the real root(s) of the cubic $x^3 + 3x^2 - x + 4 = 0$ by first replacing x by $(X - 1)$ and solving the cubic in X.

Exercise 10.5   (much harder)   Van der Waals' equation for 1 mole of a gas is

$$p = \frac{RT}{V - b} - \frac{a}{V^2}$$

This rearranges to be a cubic in V:

$$V^3 - (b + RT/p)V^2 + (a/p)V - ab/p = 0.$$

R is the gas constant, a and b are constants which depend on the gas under consideration and p, V, T are the pressure, volume and temperature of the gas. The van der Waals' equation, being cubic, suggests that there may be more than one possible volume for a given p and T.

Investigate whether this is really so for carbon dioxide gas:

$$R = 8 \cdot 314 \text{ J mol}^{-1} \text{ K}^{-1}$$
$$a = 0 \cdot 3636 \text{ Pa m}^6 \text{ mol}^{-2}$$
$$b = 4 \cdot 267 \times 10^{-5} \text{ m}^3 \text{ mol}^{-1}$$

Take $p = 50 \times 10^5$ Pa (≈ 51 atmospheres) and

(i) $T = 280$ K;

(ii) $T = 500$ K.

(Hint: having set up the cubic in V make a change of scale by letting $v = V \times 10^4$. This makes the coefficients in the cubic much more manageable.)

NUMERICAL METHODS OF SOLVING NON-LINEAR EQUATIONS

METHOD OF BISECTION

To find the root of a continuous function f we require to find an x such that $f(x) = 0$. In graphical terms we must find where the graph meets the x-axis. If we can find two values $x = a$ and $x = b$ such that $f(a)$ and $f(b)$ are of opposite sign then we know that there is at least one root $x = \alpha$ somewhere between a and b (fig. 10.1).

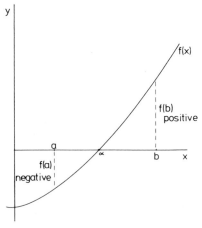

Fig. 10.1

This leads us to a very simple but slowly converging method for locating a root of $f(x)$, called the Method of Bisection.

The first stage is to find two points a, b on opposite sides of the root. Then the function's value at the midpoint between a and b is computed, i.e. $f[(a+b)/2]$. Three cases arise:

   (i) $f(a)$ and $f[(a+b)/2]$ are of opposite sign

   (ii) $f[(a+b)/2]$ is zero

   (iii) $f(b)$ and $f[(a+b)/2]$ are of opposite sign.

In case (i) we now know that the root lies between $x=a$ and $x=(a+b)/2$ and so have halved the interval in which it is known to lie.

In case (ii) we have found the root, at $x=(a+b)/2$.

In case (iii) the root must lie in the interval from $x=(a+b)/2$ to $x=b$.

This process can be repeated until the root is located in a very small interval, and so known accurately (fig. 10.2).

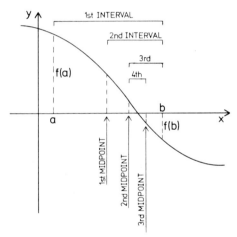

Fig. 10.2

Example 1

Find a root of $f(x)=x^5-3x^3+1=0$.

First we form a table of values, using the calculator to evaluate $f(x)$, for various x, until a change of sign is seen:

x	0	0·4	0·8
$f(x)$	1	0·818	−0·208

Clearly there is a root between 0·4 and 0·8.

We now find the value at the midpoint ($x=0·6$)

$$f(0·6)=0·42976$$

This is of the same sign as $f(0\cdot4)$ and opposite to that of $f(0\cdot8)$ so the root must lie between $0\cdot6$ and $0\cdot8$.

We now find $f(x)$ at the midpoint of $x=0\cdot6$ and $x=0\cdot8$ and again compare:

$f(0\cdot7)=0\cdot13907$. This is again positive and of opposite sign to $f(0\cdot8)$ so the root lies in the interval $x=0\cdot7$ to $x=0\cdot8$.

Continuing: $f(0\cdot75)=-0\cdot02832$. We have now gone just past the root since $f(0\cdot75)$ and $f(0\cdot8)$ are both negative so the root lies between $0\cdot7$ and $0\cdot75$.

This process can be illustrated graphically (fig. 10.3).

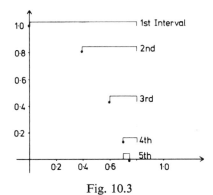

Fig. 10.3

Although rather slow and tedious, the process is more-or-less fool-proof and bound to converge to the root. However, the method cannot be used to find an *even multiple* root where the curve just touches the x-axis and is of the same sign on both sides.

It is helpful to set out the working in tabular form, as illustrated by the example below.

Example 2

SO_2 and O_2 react reversibly to form SO_3:

$$2SO_2+O_2\rightleftharpoons2SO_3$$

Let a be the initial concentration of O_2 and b the initial concentration of SO_2. If x is the number of mol l^{-1} of O_2 consumed, then at equilibrium:

$$\frac{[SO_3]^2}{[O_2][SO_2]^2}=\text{equilibrium constant}=\frac{(2x)^2}{(a-x)(b-2x)^2}$$

Problem

At 900°C the equilibrium constant$=0\cdot13$. If the starting amount of O_2 is $4\cdot3$ mol l^{-1} and of SO_2 is $6\cdot0$ mol l^{-1} find the equilibrium concentrations of the three substances.

Solution

We must solve the cubic equation

$$\frac{4x^2}{(4\cdot3-x)(6-2x)^2}=0\cdot13$$

Multiplying across: $4x^2=0\cdot13(4\cdot3-x)(6-2x)^2$

Multiplying out: $4x^2=20\cdot124-18\cdot096x+5\cdot356x^2-0\cdot52x^3$

Rearranging: $0\cdot52x^3-1\cdot356x^2+18\cdot096x-20\cdot124=0$

Dividing by $0\cdot52$: $x^3-2\cdot6076923x^2+34\cdot8x-38\cdot7=0$

We now evaluate the cubic, $f(x)$, at a few simple points to locate a real root:

x	0	1	1·4
$f(x)$	$-38\cdot7$	$-5\cdot51$	$+7\cdot65$

The change of sign of $f(x)$ indicates a root between $x=1$ and $x=1\cdot4$. We set out the results of our bisection method calculations in Table 10.1. The $f(x)$ function evaluations are performed on a calculator using nested multiplication:

$$[(x-2\cdot6076923)\times x+34\cdot8]\times x-38\cdot7.$$

TABLE 10.1

	Interval	New x value	$f(x)$ value	New x will replace x on line:
(a)		1	$-5\cdot51$	
(b)		1·4	$+7\cdot65$	
(c)	[1, 1·4]	1·2	$+1\cdot03$	(b)
(d)	[1, 1·2]	1·1	$-3\cdot58$	(a)
(e)	[1·1, 1·2]	1·15	$-0\cdot61$	(d)
(f)	[1·15, 1·2]	1·175	$+0\cdot22$	(c)
(g)	[1·15, 1·175]	1·1625	$-0\cdot20$	(e)
(h)	[1·1625, 1·175]	1·16875	$+0\cdot007$	(f)
(i)	[1·1625, 1·16875]	1·165625	$-0\cdot096$	(g)
(j)	[1·165625, 1·16875]	1·1671875	$-0\cdot044$	(i)
(k)	[1·1671875, 1·16875]	1·16796875	$-0\cdot019$	(j)
(l)	[1·16796875, 1·16875]	1·168359375	$-0\cdot006$	(k)
(m)	[1·168359375, 1·16875]			

At this stage it is clear that the root is at about 1·168. To see whether the answer to be quoted is 1·168 or 1·169 to 3 d.p. we evaluate the cubic at $1\cdot1685$, which gives $f(1\cdot1685) = -0\cdot001$, so the root is between $1\cdot1685$ and $1\cdot16875$. In fact this tells us that the root is $1\cdot169$ to 3 d.p. [*Note:* The above procedure of precisely halving the interval is somewhat tedious and really of no advantage. The long string of decimal digits can be avoided by rounding off values and so dealing with slightly different intervals.]

We have now found $x = 1\cdot169$ so the equilibrium concentrations are:

$$SO_3: 2x = 2\cdot34; \quad O_2: a-x = 3\cdot13; \quad SO_2: b-2x = 3\cdot66$$

Some things to do

The Exercises numbered 10.1 to 10.5 and 10.12 to 10.18 are all suitable for solution by the Method of Bisection. The reader can select a few to try.

LINEAR ITERATION

Often a simple recurrence relation can be found which produces a sequence of numbers converging to the solution of an equation. Such a scheme is Linear Iteration where an estimate of the root is found and used to derive a better estimate ... and so on. The equation in x, $f(x) = 0$, is rearranged to the form $x = F(x)$ and then the iterative formula $x_{n+1} = F(x_n)$ used.

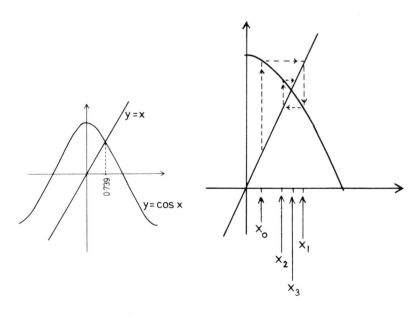

Fig. 10.4   Fig. 10.5

The following 'party trick' illustrates the method. Ask a friend to enter any decimal number between 0.0 and 1.0 (the actual range allowed is that for the $\boxed{\cos}$ function) into the display of a calculator. Tell him to select radian mode and then press the $\boxed{\cos}$ function key 30 times and then round off the number displayed to 4 d.p. His answer should be 0.7391 *whatever the starting value*. What he is really doing is solving the equation $x = \cos x$ by linear iteration. Suppose that 0.29 is entered. This means $x_0 = 0.29$. Now $\boxed{\cos}$ is pressed, which produces $x_1 = \cos x_0 = \cos 0.29 = 0.95824388$. Now $\boxed{\cos}$ is pressed again, which produces $x_2 = \cos x_1 = \cos 0.95824388 = 0.57495770$.

This process continues and the displayed values converge towards the root 0.73908513. (Try this for yourself.)

The procedure can be demonstrated graphically (figs. 10.4 and 10.5).

We shall now present a more detailed description of the method of linear iteration.

Example

Find the roots of $x^2 - 3x + 1.25 = 0$.

Solution using an Iterative Formula

We take $x^2 - 3x + 1.25 = 0$ and rearrange it into the form $x = F(x)$:

$$x^2 = 3x - 1.25$$

$$\therefore \quad x = \sqrt{(3x - 1.25)}$$

We can take as our Iterative Formula

$$x_{n+1} = \sqrt{(3x_n - 1.25)}$$

If we take as our first estimate $x_0 = 2$ then

$$x_1 = \sqrt{(3x_0 - 1.25)} = \sqrt{(3 \times 2 - 1.25)} = 2.179449472$$

$$x_2 = \sqrt{(3x_1 - 1.25)} = \sqrt{(3 \times 2.179449472 - 1.25)} = 2.299640932,$$

and so on, working with a 10-digit calculator.

Table 10.2 shows the sequence generated.

TABLE 10.2

Term	Value		Term	Value
x_0	$2.$		x_7	2.483489205
x_1	2.179449472		x_8	2.490073817
x_2	2.299640932		x_9	2.494037179
x_3	2.376746262		x_{10}	2.496419744
x_4	2.424920367		x_{11}	2.497850923
x_5	2.454538878		x_{12}	2.498710221
x_6	2.472572877		x_{13}	2.499226013

This is clearly converging (rather slowly) to the true root $x = 2 \cdot 5$.

The calculator can be used most effectively with such an iterative problem. No memory is needed and the whole process can be performed 'automatically' once the initial estimate (x_0) is entered.

	Algebraic *key strokes*	*Display*	*Comment*	*Reverse Polish* *key strokes*
	2	2	x_0	2 [enter]
Cycle 1	[×] 3 [−]			3 [×] 1·25
	1·25 [=]	4·75	$3x_0 - 1 \cdot 25$	[−]
	[√x]	2·179449472	x_1	[√x]
Cycle 2	[×] 3 [−]			3 [×] 1·25
	1.25 [=]	5·288348415	$3x_1 - 1 \cdot 25$	[−]
	[√x]	2·299640932	x_2	[√x]
	and so on.			and so on.

There is really no need to record the intermediate results (as was done in Table 10.2) and the iterative process is thereby speeded up considerably. The key-stroke sequence (Algebraic) [×] 3 [−] 1·25 [=] [√x] *or* (Reverse Polish) 3 [×] 1·25 [−] [√x] is repeated until no significant change is occurring in the successive estimates displayed.

If a memory is available the 1·25 can be initially stored and recalled each time, thereby saving key strokes. There is no saving by putting the 3 into memory as it is a 1-digit number but if the iterative formula had been

$$x_{n+1} = \sqrt{(3 \cdot 49213 x_n - 1 \cdot 25647)}$$

clearly it would be helpful to be able to store both the constants, to save key strokes and reduce the chance of errors. However it is worth noting that usually iterative processes are self-correcting. If an error is made the sequence of estimates will still ultimately converge or diverge just as if no error had been made.

A second Iterative Formula solution

In the section above we took the equation $x^2 - 3x + 1 \cdot 25 = 0$ and rearranged it into the required form $x = F(x)$, to give $x = \sqrt{(3x - 1 \cdot 25)}$.

There are several other ways of rearranging the original equation. For example:

$$x^2 - 3x + 1 \cdot 25 = 0$$

$$\therefore \quad 3x = x^2 + 1 \cdot 25$$

$$\therefore \quad x = \tfrac{1}{3}(x^2 + 1 \cdot 25)$$

This is quite different and so we have an alternative Iterative Formula from the same initial equation, namely:

$$x_{n+1} = \tfrac{1}{3}(x_n^2 + 1 \cdot 25)$$

Taking $x_0 = 2$ again, Table 10.3 shows the sequence produced using an 8-digit calculator.

TABLE 10.3

Term	Value	Term	Value
x_0	2·	x_7	0·5191509
x_1	1·75	x_8	0·5065059
x_2	1·4375	x_9	0·5021827
x_3	1·1054687	x_{10}	0·5007291
x_4	0·8240203	x_{11}	0·5002432
x_5	0·6430031	x_{12}	0·5000810
x_6	0·5544843	x_{13}	0·5000027

This is clearly converging (and a little more rapidly) but this time to the *other* root of the quadratic equation. Once again the calculator can be used to perform all the calculations 'automatically':

	Algebraic key strokes	*Display*	*Comment*	*Reverse Polish key strokes*
	2	2	x_0	2
Cycle 1	$\boxed{x^2}$ $\boxed{+}$ 1·25 $\boxed{\div}$ 3 $\boxed{=}$	1·75	x_1	$\boxed{x^2}$ 1·25 $\boxed{+}$ 3 $\boxed{\div}$
Cycle 2	$\boxed{x^2}$ $\boxed{+}$ 1·25 $\boxed{\div}$ 3 $\boxed{=}$	1·4375	x_2	$\boxed{x^2}$ 1·25 $\boxed{+}$ 3 $\boxed{\div}$
	and so on.			and so on.

Divergence

Does it matter what we take as the initial estimate x_0?

Suppose we use $x_{n+1} = \tfrac{1}{3}(x_n^2 + 1 \cdot 25)$ with $x_0 = 2 \cdot 6$, which is quite close to the larger root (2·5). The following sequence is generated (Table 10.4).

TABLE 10.4

Term	Values	Term	Values
x_0	2·6	x_7	15·95 . . . . .
x_1	2·67	x_8	85·25 . . . . .
x_2	2·79	x_9	2423·3 . . . .
x_3	3·01	x_{10}	$1 \cdot 9 .. \times 10^6$ ..
x_4	3·45	x_{11}	$1 \cdot 2 .. \times 10^{12}$.
x_5	4·38	x_{12}	$5 \cdot 4 .. \times 10^{23}$.
x_6	6·82	x_{13}	$9 \cdot 8 .. \times 10^{46}$

Obviously this is diverging and not tending towards the value of either of the roots (0·5 or 2·5).

However $x_0 = 2·6$ is quite satisfactory when used with the first iterative formula $x_{n+1} = \sqrt{(3x_n - 1·25)}$, and the sequence then converges to 2·5. (The interested reader can check this.)

It must be concluded that whether an iterative formula will converge or diverge sometimes depends on the starting value chosen and also on the particular formula itself. Usually trial-and-error will be adequate in finding a suitable formula and starting value. The theory behind the method can be found in standard books on numerical methods and is indicated in Exercise 10.11.

Rate of Convergence

When linear iteration methods work the 'errors' (differences between the estimates and the true answer) usually decrease by a roughly constant factor each time when getting near to the root. For example, the formula $x_{n+1} = \frac{1}{3}(x_n^2 + 1·25)$ starting with $x_0 = 2$ (Table 10.3) produces the results in Table 10.5.

TABLE 10.5

Term	Approx. Error	Factor	Term	Approx. Error	Factor
x_0	1·5	—	x_7	0·01915097	·345
x_1	1·25	·833	x_8	0·00650591	·336
x_2	0·9375	·750	x_9	0·00218275	·334
x_3	0·6055	·646	x_{10}	0·00072917	·334
x_4	0·3240	·535	x_{11}	0·00024323	·334
x_5	0·1430	·441	x_{12}	0·00008110	·333
x_6	0·0555	·388	x_{13}	0·00002703	·333

We see that the factor by which an error is multiplied to give the next error settles down at 0·333.

So: $\text{error}_{n+1} = 0·333 \times \text{error}_n$.

This is a *linear* relationship between the errors, hence the name *linear iteration* for the method.

Some things to do

Exercise 10.6   The illustrative example in the text, $x^2 - 3x + 1·25 = 0$, can be rearranged as follows:

$$x = \frac{-1·25}{x - 3}$$

So another iterative formula is:

$$x_{n+1} = \frac{-1 \cdot 25}{x_n - 3}$$

By taking various starting values investigate the convergence/ divergence of this formula. (When using a calculator the reciprocal key $\boxed{1/x}$ is very useful here. Given x_n, in the display, compute $(x_n - 3)/-1 \cdot 25$ and then take the reciprocal to give x_{n+1}. Keeping $-1 \cdot 25$ stored in a memory makes the procedure even simpler.)

Exercise 10.7 Investigate whether or not $x_{n+1} = \frac{1}{3}(x_n^2 + 1 \cdot 25)$ can be used to converge to $x = 2 \cdot 5$ with a suitable choice of x_0.

Exercise 10.8 It was shown in the text that by entering any value and repeatedly pressing $\boxed{\cos}$ we found a root of $x = \cos x$.

(a) Try doing the same for other function keys

  (i) $\boxed{\sin}$ (radians); (ii) $\boxed{\sin}$ (degrees); (iii) $\boxed{e^x}$; (iv) $\boxed{\ln}$ or $\boxed{\log}$; (v) $\boxed{\tan}$ (radians); (vi) $\boxed{\tan}$ (degrees).

(b) What equations are you really trying to solve? If the calculator settles down to a constant value is it a root of the equation? If the values get too large, then no root can be found by this method. Is this because no root exists?

(c) What happens if $\boxed{\sqrt{x}}$ is pressed repeatedly, starting from any non-negative number? What equation is solved? What are the roots?

(d) Repeat (c) for $\boxed{x^2}$.

(e) If your calculator has them, try the inverse functions and the hyperbolic functions.

Exercise 10.9 The equation

$$\frac{103}{x} - \frac{161}{1-x} - \frac{515}{2-x} + \frac{461}{1+x} = 0$$

occurs in a genetics problem. Rearranging to the form

$$x = \frac{130(x^3 - 2x^2 - x + 2)}{815x^2 - 1222x + 729}$$

yields an iterative scheme. Taking $x_0 = 0$ find a root of this equation.

Exercise 10.10 For the iterative procedure $x_{n+1} = \sqrt{(3x_n - 1 \cdot 25)}$, $x_0 = 2$, the sequence of estimates converges to $2 \cdot 5$. This was discussed in the text and the results presented in Table 10.2 (page 177). Show that the errors decrease by a more-or-less constant factor, setting out the results as in Table 10.5.

Exercise 10.11 It is shown in numerical analysis books that for the linear iteration formula

$$x_{n+1} = F(x_n)$$

the successive errors e_n, and e_{n+1} are approximately related by

$$e_{n+1} = ke_n$$

where $k = dF(x)/dx$ evaluated at $x = $ root.
   For example, the formula in Exercise 10.10 has $F(x) = \sqrt{(3x - 1 \cdot 25)}$
so

$$\frac{dF(x)}{dx} = \frac{\frac{1}{2} \times 3}{\sqrt{(3x - 1 \cdot 25)}}.$$

A root is at $x = 2 \cdot 5$ so

$$\frac{dF(x)}{dx} = \frac{\frac{1}{2} \times 3}{\sqrt{(3 \times 2 \cdot 5 - 1 \cdot 25)}} = 0 \cdot 6$$

which is the factor found experimentally. Only if $F'(x)$ is numerically smaller than 1 can convergence be expected.
   Analyse the convergence of the iterative formulae below when the estimate is near to the root

(a) $x_{n+1} = \frac{1}{3}(x_n^2 + 1 \cdot 25)$; root $0 \cdot 5$ and then try root $2 \cdot 5$

(b) $x_{n+1} = \cos x_n$; root $0 \cdot 739$

(c) $x_{n+1} = \sqrt{(3x_n - 1 \cdot 25)}$; root $0 \cdot 5$

(d) $x_{n+1} = \dfrac{-1 \cdot 25}{x_n - 3}$; root $0 \cdot 5$ and root $2 \cdot 5$

(e) $x_{n+1} = \sin x_n$; root 0

(f) $x_{n+1} = \tan x_n$; root 0.

   All these iterative formulae have arisen as practical problems in this chapter. Do the theoretical convergence/divergence forecasts match reality?

NEWTON'S METHOD

This iterative method, also known as the Newton–Raphson method, is a widely used and generally very satisfactory method. However, it does require the differentiation of the function $f(x)$. The principles of the method were established by Isaac Newton in 1671 but Joseph Raphson, a contemporary of Newton, published a simpler and better version in 1690, which is the one normally used. The method is important for several reasons. It was the first to make use of calculus techniques, it is relatively simple, and it is very powerful—i.e. it gives accurate answers very quickly for a wide variety of problems.

Only one initial estimate is needed, $x=x_0$ at point P_0 say, then the tangent line to the curve through P_0 is extended to cut the x axis. This should give a better estimate, x_1 (fig. 10.6). This process is continued: P_1—the point $(x_1, f(x_1))$ is located and the tangent through P_1 is drawn to cut the x axis at x_2, and so on.

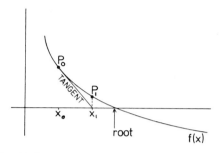

Fig. 10.6   Newton's method of root location.

The iterative formula for this process is shown by recourse to co-ordinate geometry to be:

$$x_{n+1} = x_n - \frac{f(x_n)}{f'(x_n)}$$

Note that $f'(x)$, the derivative of $f(x)$, is needed. This is so because the gradient of the tangent to the curve $f(x)$ is $f'(x)$.

Example 1

Find a root of $f(x) = x^2 - 3x + 1\cdot25 = 0$.

This example—previously solved by Linear Iteration (page 177)—well illustrates the power of Newton's Method:

$$f'(x) = 2x - 3,$$

so the iterative formula

$$x_{n+1} = x_n - \frac{f(x_n)}{f'(x_n)}$$

is

$$x_{n+1} = x_n - \frac{x_n^2 - 3x_n + 1\cdot25}{2x_n - 3}$$

We can reformulate this to the more convenient form:

$$x_{n+1} = \frac{x_n^2 - 1\cdot25}{2x_n - 3}$$

Now taking

$$x_0 = 2$$

we commence calculating:

$$x_1 = \frac{2^2 - 1 \cdot 25}{2 \times 2 - 3} = 2 \cdot 75$$

$$x_2 = \frac{(2 \cdot 75)^2 - 1 \cdot 25}{2 \times 2 \cdot 75 - 3} = 2 \cdot 525$$

$$x_3 = \frac{(2 \cdot 525)^2 - 1 \cdot 25}{2 \times 2 \cdot 525 - 3} \approx 2 \cdot 500305$$

$$x_4 = \frac{(2 \cdot 500305)^2 - 1 \cdot 25}{2 \times 2 \cdot 500305 - 3} \approx 2 \cdot 500000046$$

The convergence, once near the root, is very rapid, as is generally the case with Newton's Method. If the initial estimate is a long way from the root Newton's Method may diverge in some cases, and the root not be found.

Example 2

The finding of the smallest positive root of $f(x) - \tan(x) - x - 0 \cdot 02 = 0$, will be attempted using Newton's Method.

$$f(x) = \tan(x) - x - 0 \cdot 02$$

$$\therefore \quad f'(x) = \sec^2(x) - 1 = \tan^2(x).$$

$$x_{n+1} = x_n - \frac{f(x_n)}{f'(x_n)}$$

in this case is

$$x_{n+1} = x_n - \frac{\tan(x_n) - x_n - 0 \cdot 02}{\tan^2(x_n)}$$

$$x_0 = 0 \cdot 3$$
$$x_1 \approx 0 \cdot 3 - (-0 \cdot 11144186) = 0 \cdot 41144186$$
$$x_2 \approx 0 \cdot 41144186 - 0 \cdot 02576 = 0 \cdot 38568189$$
$$x_3 \approx 0 \cdot 38568189 - 0 \cdot 00202739 = 0 \cdot 3836545$$
$$x_4 \approx 0 \cdot 3836545 - 0 \cdot 0000119 = 0 \cdot 3836426$$
$$x_5 \approx 0 \cdot 3836426 - 0 \cdot 00000000 = 0 \cdot 3836426$$

Setting out the working in this fashion clearly shows that the 'correction term', $-[f(x_n)/f'(x_n)]$ is getting very small, as it should do, as we converge to the root.

The whole process is easily performed on the calculator if one memory (or the stack) is used to store the last estimate for the root.

Algebraic key strokes	Display	Comment	Reverse Polish key strokes
Select radian mode			Select radian mode
·3	·3	x_0	·3
STO			enter enter
tan	·30933625	$\tan(x_0)$	enter tan
− RCL −			x⇌y − ·02
·02 ÷	− ·01066375	$f(x_0)$	−
RCL tan x²			x⇌y tan x²
=	− ·11144186	$f(x_0)/f'(x_0)$	÷
+/− + RCL			
=	·41144186	x_1	−

(Left margin: Repetitive Cycle) (Right margin: Repetitive Cycle)

Example 3

The equation $f(x)=x^6-27x^5+105x^4-140x^3+81x^2-21x+2=0$ has arisen in connection with dynamic forces in aircraft engines. Find the largest root.

If x is large (~ 5 is the minimum) then x^6-27x^5 should be larger than the rest of the terms in the equation so *approximately* $x^6-27x^5=0$, i.e. $x\approx 0$ or $x\approx 27$. This suggests that a sensible initial estimate of the root is $x_0=27$.

Using Newton's Method we need

$$f'(x)=6x^5-135x^4+420x^3-420x^2+162x-21$$

$$x_{n+1}=x_n-\left\{\frac{x_n^6-27x_n^5+105x_n^4-140x_n^3+81x_n^2-21x_n+2}{6x_n^5-135x_n^4+420x_n^3-420x_n^2+162x_n-21}\right\}$$

This is a clear example where nested multiplication is of great value as we have two polynomials to evaluate each involving several terms. First we might prefer to reformulate the iterative formula:

$$x_{n+1}=\frac{5x_n^6-108x_n^5+315x_n^4-280x_n^3+81x_n^2-2}{6x_n^5-135x_n^4+420x_n^3-420x_n^2+162x_n-21}$$

$$x_0=27$$

A memory is very helpful for storing the current x_n which is repeatedly needed. Also it is helpful to be able to store the denominator while the numerator is being evaluated.

Successive estimates are:

24·620134, 23·216143, 22·694817, 22·626927, 22·625853, 22·625852,

and the root is seen to be 22·62585 to 7 s.f.

Note 1

In evaluating the numerator the x_n term is missing so the nested multiplication appears as $((((5x_n-108)x_n+315)x_n-280)x_n+81)x_n^2-2$. The denominator is $((((6x_n-135)x_n+420)x_n-420)x_n+162)x_n-21$.

Note 2

Some of the terms make negligible contribution and the process can therefore be simplified if some preliminary analysis is made.

Some things to do

Exercise 10.12   Find both roots of $2x^2-7x+5=0$ by Newton's Method and check by some other means.

Exercise 10.13   The equation $\sin \omega t = e^{-\alpha t}$ arises when studying the motion of a planetary gear system as used in automatic transmission systems.

For the case $\omega = 0\cdot612$ and $\alpha = 0\cdot014$ find the smallest root, t.

Exercise 10.14   A solid sphere of radius r and density d will sink in water to a depth x where

$$x^3 - 3rx^2 + 4r^3d - 0.$$

Solve this cubic equation for

(a) $r=1$, $d=0\cdot64$

(b) $r=4$, $d=1\cdot25$

and consider the physical meaning of the answers obtained.

Exercise 10.15   A mathematical model of the spread of a rumour in a village, given by A. Engel (1973), leads to the formula

$$n \ln (p/n) - 2p + 2n + 1 = 0$$

where $n+1$ inhabitants live in the village and p is the number of people who do not get to hear the rumour at all. The main assumption is that a person who has heard the rumour passes it on but stops telling others as soon as he finds someone who has already heard it. The equation reduces to

$$\ln (x) - 2x + 2 + \frac{1}{n} = 0, \quad \text{where } x = \frac{p}{n}$$

Solve this for the case $n=1000$ and so find the % of inhabitants who fail to hear the rumour. How does the % vary with n? (A graph would show this.)

Exercise 10.16   When light is passed through a narrow slit the intensity $I(x)$ varies with a parameter x according to

$$I(x) = \frac{\sin^2 x}{x^2}$$

(a) Find a solution of $I(x)=0$.

(b) When is $I(x)$ a maximum?

Exercise 10.17 The brachistochrone problem, discussed in Exercise 6.17 page 105 seeks to establish the curve down which a body should fall under gravity from point O to point P to get from O to P in the shortest possible time. It was shown by Newton that the cycloid, $x=r(\theta-\sin\theta)$, $y=r(1-\cos\theta)$ is the best path. To compare times for the cycloid and a straight path it is necessary to determine the point P where a line $y=kx$ will cut the cycloid (see fig. 10.7).

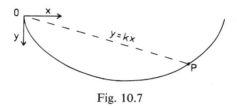

Fig. 10.7

This means solving

$$k=\frac{1-\cos\theta}{\theta-\sin\theta}$$

Use a numerical method to find θ when $k=5$.

Exercise 10.18 *The Problem* Less and Moore are next door neighbours and both decide to paint their houses at the same time. They set up their ladders as indicated in the diagram just clearing the 2 m fence which divides their properties. Less's ladder is 6 m long and Moore's is 8 m long. How far apart are the houses?

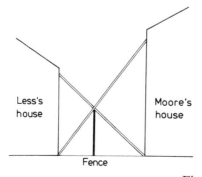

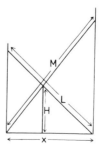

Fig. 10.8

The Mathematics It can be shown using Pythagoras' theorem that for ladders of lengths L and M crossing at a height H (see fig. 10.8) the distance apart of their feet, x, is given by

$$\frac{1}{H} = \frac{1}{\sqrt{(L^2 - x^2)}} + \frac{1}{\sqrt{(M^2 - x^2)}}$$

(This is left as an optional exercise.)

The Computation

In the example $H = 2$, $L = 6$, $M = 8$ so we must solve

$$\tfrac{1}{2} = \frac{1}{\sqrt{(36 - x^2)}} + \frac{1}{\sqrt{(64 - x^2)}},$$

i.e.

$$f(x) = \tfrac{1}{2} - \frac{1}{\sqrt{(36 - x^2)}} - \frac{1}{\sqrt{(64 - x^2)}} = 0.$$

To attempt this by algebra leads to a quartic equation so a numerical method is advisable. The choice is yours!

Exercise 10.19   The equation $\tan x + \tanh x = 0$ occurs in the theory of vibrations. Find the smallest positive root.

11. Simultaneous linear equations

For a pair of simultaneous equations such as

$$5x + 3y = 11 \ldots \text{(i)}$$
$$3x - 2y = -1 \ldots \text{(ii)}$$

a common method of solution is:

$2 \times$ (i)	$10x + 6y = 22$	
$3 \times$ (ii)	$9x - 6y = -3$	
add	$19x \quad = 19$	(This eliminates y)
divide by 19	$x = 1$	

substitute for x

back in (i)	$5 + 3y = 11$	(Having found x, y is now sought)
subtract 5	$3y = 6$	
divide by 3	$y = 2$	

$$\therefore \quad \text{Solution is } x = 1, \ y = 2$$

Similar processes can be devised for more complicated problems: 3 equations in 3 unknowns, 4 equations in 4 unknowns. . . . However with large sets of simultaneous linear equations, two problems arise. Firstly the 'by inspection' method of eliminating the variables is laborious and a systematic method is needed. Secondly the exact arithmetic becomes unwieldy—one must carry either very large numbers or very many decimal places—or one must approximate and so risk getting an inexact solution.

The pocket calculator proves to be a very valuable aid when solving sets of equations (up to about 6 unknowns can be reasonably handled) and for larger sets (maybe 100 unknowns!) a computer is rather essential. Simultaneous equations arise in electrical circuit analysis, structural engineering, and indeed may be come across throughout science.

METHOD OF GAUSSIAN ELIMINATION

This is basically a systematized form of the simple method already mentioned, and is best illustrated by an example.

For the electronic circuit whose diagram is shown in fig. 11.1 it is required to calculate the currents i_1, i_2, i_3 which will flow when the connection is made.

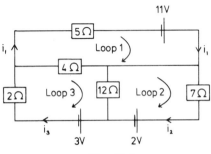

Fig. 11.1

Using the relationship $V = IR$ for the three loops of the circuit we have:

Loop 1: $5i_1 + 4(i_1 - i_3) = 11$
Loop 2: $7i_2 + 12(i_2 - i_3) = -2$
Loop 3: $2i_3 + 4(i_3 - i_1) + 12(i_3 - i_2) = -3$

Simplifying:

$$9i_1 \qquad\quad - 4i_3 = 11$$
$$19i_2 - 12i_3 = -2$$
$$-4i_1 - 12i_2 + 18i_3 = -3$$

The aim is to modify these equations systematically by adding multiples of each other together to produce three equations in the form

$$Ai_1 + Bi_2 + Ci_3 = K \quad \ldots \quad (a)$$
$$Di_2 + Ei_3 = L \quad \ldots \quad (b)$$
$$Fi_3 = M \quad \ldots \quad (c)$$

where $A, B, C, D, E, F, K, L, M$ are known numbers.

Having obtained this form i_3 is obtained from (c) since $i_3 = M/F$. Then this known value for i_3 is substituted back into (b):

$$Di_2 = L - Ei_3 \quad \text{so} \quad i_2 = \frac{L - Ei_3}{D}$$

and finally the values of i_2 and i_3 are substituted back into (a):

$$Ai_1 = K - Bi_2 - Ci_3$$

so

$$i_1 = \frac{K - Bi_2 - Ci_3}{A}$$

Example:

$$9i_1 \qquad\quad - 4i_3 = 11 \quad \dots \text{ (a)}$$
$$19i_2 - 12i_3 = -2 \quad \dots \text{ (b)}$$
$$-4i_1 - 12i_2 + 18i_3 = -3 \quad \dots \text{ (c)}$$

First we eliminate i_1 from rows (b) and (c). In fact i_1 is already absent from row (b) so we simply add 4/9 of row (a) to row (c):

$$9i_1 \qquad\qquad - 4i_3 = 11 \qquad\qquad \dots \text{ (a)}$$
$$19i_2 \qquad - 12i_3 = -2 \qquad\qquad \dots \text{ (b)}$$
$$[c + \tfrac{4}{9}a] \qquad - 12i_2 + 16 \cdot 222222i_3 = 1 \cdot 8888888 \quad \dots \text{ (d)}$$

It now remains to eliminate i_2 from row (d), done by adding 12/19 of row (b) to row (d):

$$9i_1 \qquad\qquad - 4i_3 = 11 \qquad\qquad \dots \text{ (a)}$$
$$19i_2 - 12i_3 = -2 \qquad\qquad \dots \text{ (b)}$$
$$[d + \tfrac{12}{19}b] \qquad 8 \cdot 6432749i_3 = 0 \cdot 625731 \quad \dots \text{ (e)}$$

The three equations are now:

$$9i_1 \qquad\qquad - 4i_3 = 11 \qquad\qquad \dots \text{ (a)}$$
$$19i_2 \qquad - 12i_3 = -2 \qquad\qquad \dots \text{ (b)}$$
$$8 \cdot 6432749i_3 = 0 \cdot 625731 \quad \dots \text{ (e)}$$

We can now divide (e) by 8·6432749 to get

$$[e/8 \cdot 6432749] \qquad i_3 = \frac{0 \cdot 625731}{8 \cdot 6432749} = 0 \cdot 072395128$$

and now back substitute in (b):

$$i_2 = \frac{12i_3 - 2}{19} = \frac{12 \times 0 \cdot 072395128 - 2}{19}$$
$$= -0 \cdot 059539919$$

and in (a):

$$i_1 = \frac{11 + 4i_3}{9} = 1 \cdot 2543978$$

As a measure of the accuracy of these results

$$i_1 = 1 \cdot 2543978, \quad i_2 = -0 \cdot 059539919, \quad i_3 = 0 \cdot 072395128$$

we calculate the *residuals*. These residuals will be zero for an exact solution, and are the original equations rewritten to have zero on the right-hand-side.

$$R_1 = 9i_1 - 4i_3 - 11$$
$$R_2 = 19i_2 - 12i_3 + 2$$
$$R_3 = -4i_1 - 12i_2 + 18i_3 + 3$$

For our solution set we get:

$$R_1 \approx -3 \cdot 1 \times 10^{-6}$$
$$R_2 \approx 1 \cdot 6 \times 10^{-6}$$
$$R_3 \approx 1 \cdot 3 \times 10^{-6}$$

These are clearly very small residuals suggesting that the solution is quite accurate. (Actually it is possible to have very small residuals but a very inaccurate solution in cases when the equations are 'ill-conditioned' but we shall not go into this.)

If the original equations are solved *exactly* we get:

$$i_1 = \frac{8343}{6651}$$

$$i_2 = \frac{-836}{14041}$$

$$i_3 = \frac{107}{1478}$$

The reader can compare these with the results obtained above. Using the pocket calculator to solve simultaneous equations is quite straightforward. As there is repeated need to make calculations of the type

$$1 \times 2 + 3 \times 4 + 5 \times 6 + \ldots$$

it is obvious that 'sum of products' logic or brackets (now becoming widely available) or Reverse Polish logic are a very great help. Otherwise each product must be formed separately, and preferably a memory be used to accumulate the sum of the products.

It is advisable to set out the working in tabular form. This is illustrated (Table 11.1) for the circuit problem solved above to which the reader should refer.

TABLE 11.1

Row	Formation	i_1	i_2	i_3	R.H.S.	Comment
(a)	—	9	0	−4	11	First equation
(b)	—	0	19	−12	−2	Second equation
(c)	—	−4	−12	18	−3	Third equation
(d)	$c + \frac{4}{9}a$	0	−12	16·222222	1·8888888	Eliminate i_1 from (c)
(e)	$d + \frac{12}{19}b$	0	0	8·6432749	0·625731	Eliminate i_2 from (d)
(f)	$e \div 8·6432749$	0	0	1	0·072395128	Solution i_3
(g)	$(b + 12f) \div 19$	0	1	0	−0·059539919	Solution i_2
(h)	$(a + 4f) \div 9$	1	0	0	1·2543978	Solution i_1

GAUSS–SEIDEL ITERATIVE METHOD

This essentially very different method is particularly useful when certain numbers in the simultaneous equations are much larger than the others.

If a set of n simultaneous equations can be rearranged as follows:

$$\underline{a_1}x_1 + a_2x_2 + a_3x_3 + \ldots + a_nx_n = A$$
$$b_1x_1 + \underline{b_2}x_2 + b_3x_3 + \ldots + b_nx_n = B$$
$$c_1x_1 + c_2x_2 + \underline{c_3}x_3 + \ldots + c_nx_n = C$$
$$\cdot \quad \cdot \quad \cdot \quad \cdot \quad \cdot \quad \cdot \quad \cdot \quad \cdot \quad \cdot$$
$$r_1x_1 + r_2x_2 + r_3x_3 + \ldots + \underline{r_n}x_n = R$$

such that each coefficient on the main diagonal (underlined) is larger in magnitude than the sum of the magnitudes of all the other coefficients in its particular equation (left-hand-side coefficients that is) then the method is sure to work.

Example:

$$\underline{10}x + y + 2z = 9 \qquad \text{(i)}$$
$$-2x - \underline{12}y + z = 20 \qquad \text{(ii)}$$
$$x - y + \underline{9}z = 0 \qquad \text{(iii)}$$

The underlined coefficients are large enough because

$$\text{(i)} \quad 10 > 1 + 2$$
$$\text{(ii)} \quad 12 > 2 + 1$$
$$\text{(iii)} \quad 9 > 1 + 1$$

and so the following Gauss–Seidel method is sure to work:
The equations are rearranged as follows:

$$x = \tfrac{1}{10}(9 - y - 2z) \qquad \ldots \quad \text{I}$$
$$y = \tfrac{1}{12}(-20 - 2x + z) \qquad \ldots \quad \text{II}$$
$$z = \tfrac{1}{9}(0 - x + y) \qquad \ldots \quad \text{III}$$

If exactly correct values for y ānd z are substituted on the right-hand-side of Eqn. I then its value will be the correct value of x.

If we make the initial guess that $x=x_0=0$, $y=y_0=0$, $z=z_0=0$ we can use Eqn. I to calculate a new estimate, x_1, for x:

$$x_1=\tfrac{1}{10}(9-y_0-2z_0)=\tfrac{1}{10}(9-0-0)=0\cdot9$$

We then use Eqn. II to find a similar estimate, y_1, for y making use of the newly found estimate for x (i.e. x_1):

$$y_1=\tfrac{1}{12}(-20-2x_1+z_0)=\tfrac{1}{12}(-20-1\cdot8+0)\approx-1\cdot8166667$$

$\underset{\substack{\text{new estimate} \\ \text{for } y}}{\uparrow} \qquad \underset{\substack{\text{latest} \\ \text{estimate} \\ \text{for } x}}{\uparrow} \quad \underset{\substack{\text{latest} \\ \text{estimate} \\ \text{for } z}}{\uparrow}$

Next we use Eqn. III to find a new estimate for z:

$$z_1=\tfrac{1}{9}(-x_1+y_1)\approx\tfrac{1}{9}(-0\cdot9-1\cdot8166667)\approx-0\cdot30185186$$

$\underset{\substack{\text{next} \\ \text{estimate} \\ \text{for } z}}{\uparrow} \quad \underset{\substack{\text{latest} \\ \text{estimate} \\ \text{for } x}}{\uparrow} \quad \underset{\substack{\text{latest} \\ \text{estimate} \\ \text{for } y}}{\uparrow}$

Now we have a new set of estimates x_1, y_1, z_1 and can repeat to compute x_2, y_2, z_2, and so on, which ultimately should converge to the correct solutions. Note that the latest calculated values of the variables are used in the calculations at every stage. Also it is unimportant in which order equations I, II, III are written down and used.

The three iterative equations used here can be written as:

$$x_{r+1}=\tfrac{1}{10}(9-y_r-2z_r)$$
$$y_{r+1}=\tfrac{1}{12}(-20-2x_{r+1}+z_r)$$
$$z_{r+1}=\tfrac{1}{9}(-x_{r+1}+y_{r+1})$$

TABLE 11.2

r	0	1	2	3
x_r	0	0·9	1·1420371	1·1554204
y_r	0	−1·8166667	−1·8821605	−1·8872386
z_r	0	−0·30185186	−0·33602196	−0·33807322

r	4	5	6
x_r	1·1563385	1·1563985	1·1564025
y_r	−1·8875625	−1·887584	−1·8875854
z_r	−0·33821122	−0·33822028	−0·33822088

The working can be set out systematically as shown in Table 11.2 and it is clear that the values are converging and we would be justified to quote

$$x = 1 \cdot 1564$$
$$y = -1 \cdot 8876$$
$$z = -0 \cdot 33822$$

as the solution correct to 5 s.f.

A pocket calculator makes the Gauss–Seidel method simple to apply, particularly if memories are available in which to store the latest estimates. It might appear that three memories are needed for the 3 variable process to become automatic (i.e. no re-entry of estimates needed) but in fact two will suffice, (see page 196 for the details). However with three memories life is rather simpler as one memory can be used permanently for each of x, y, z.

The Gauss–Seidel method is more straightforward than Gaussian Elimination and can be rapid. However, it does not always converge— in some cases it fails (diverges) for all possible rearrangements of the initial equations; in some cases it converges for some arrangements but not others; in some cases it converges very slowly, requiring very many iterations. It is therefore only advisable to use the Gauss–Seidel method when the diagonal coefficients are *much larger* than all the others in their respective rows.

Some things to do

Exercise 11.1   A dietician draws up the following table showing the food content per gram of white bread, cheddar cheese and tomatoes:

	Bread	*Cheese*	*Tomatoes*
Energy (kcal)	2·53	4·12	0·14
Protein (g)	0·083	0·254	0·009
Fat (g)	0·017	0·345	0·000

[*Note*: Food energy is commonly measured in 'large calories', which are in fact kilocalories.]

How should the dietician construct a diet providing exactly 2000 kcal, 80 g protein and 45 g fat per day using only these ingredients? The linear equations are set up as follows:

Let x be the number of grams of bread, y the number of grams of cheese and z the number of grams of tomatoes. Then

Energy	$2 \cdot 53x + 4 \cdot 12y + 0 \cdot 14z = 2000$
Protein	$0 \cdot 083x + 0 \cdot 254y + 0 \cdot 009z = 80$
Fat	$0 \cdot 017x + 0 \cdot 345y = 45$

Solve these equations and decide if you would like the diet!
(Source: *Manual of Nutrition*, 7th Edition, G.B. Ministry of Agriculture, Fisheries and Food. H.M.S.O. London 1970.)

	Algebraic key strokes	Display	Comment	Memory 1	Memory 2	Reverse Polish key strokes	
Initial set-up	9 ÷ 10 = STO 1 × 2 +/− 20 ÷ 12 = STO 2 − RCL 1 ÷ 9 = STO 1	·9 −1·8166667 −·30185186	x y z	·9 ·9 −·30185186	— −1·8166667 −1·8166667	9 enter 10 ÷ STO 1 2 +/− × 20 − 12 ÷ STO 2 RCL 1 − 9 ÷ STO 1	Initial set-up
Cycle which repeats with memories 1 and 2 alternating	× 2 +/− − RCL 2 + 9 ÷ 10 = STO 2 × 2 +/− + RCL 1 − 20 ÷ 12 = STO 1 − RCL 2 − 9 = ÷ STO 2	1·1420371 −1·8821605 −·33602196	x y z	−·30185186 −1·8821605 −1·8821605	1·1420371 1·1420371 −·33602196	2 +/− × RCL 2 − 9 + 10 ÷ STO 2 2 +/− × RCL 1 + 20 − 12 ÷ STO 1 RCL 2 − 9 ÷ STO 2	Cycle which repeats with memories 1 and 2 alternating
Next cycle	× 2 +/− − RCL 1 + 9 ÷ 10 = STO 1	1·1554204 ...	x ...	1·1554204 ...	−·33602196 ...	2 +/− × RCL 1 − 9 + 10 ÷ STO 1	Next cycle

Exercise 11.2   The equilibrium of a certain structural framework requires that parameters ϕ_1, ϕ_2, ϕ_3, ϕ_4 satisfy the equations

$$9932\phi_1 + 1720\phi_2 + 321\phi_3 - 1200\phi_4 = 16\cdot62$$
$$1720\phi_1 + 10920\phi_2 - 1200\phi_3 + 562\phi_4 = -8\cdot39$$
$$321\phi_1 - 1200\phi_2 + 16121\phi_3 - 109\phi_4 = 22\cdot83$$
$$-1200\phi_1 + 562\phi_2 - 109\phi_3 + 16203\phi_4 = -11\cdot09$$

Find the required values of ϕ_1, ϕ_2, ϕ_3, ϕ_4.

Exercise 11.3   A chemical plant consists of a Mixer unit where incoming ores are blended together, and a Reactor which produces the required chemical product, and a Recycling unit which takes some of the waste products from the Reactor and eventually feeds these back to the Mixer (fig. 11.2):

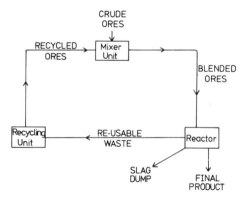

Fig. 11.2   Layout of chemical plant.

There are two main ores A and C. A is contaminated with B, and C is contaminated with D. The input rates of the crude ores are A: 300 kg per hr; B: 250 kg per hr; C: 350 kg per hr; D: 100 kg per hr. The Reactor extracts all of A and some of C from the blended ores to yield a final product which is 60% A and 40% C in composition. Slag consisting of all D is drawn off and dumped. The other waste products (consisting of the rest of C and all of B) pass into the Recycling unit which converts some of the waste products into A and D and then transmits the treated waste back into the Mixer. The treated waste is 30% A, 40% B, 20% C, 10% D.

   How much final product will the chemical plant produce per hour? To solve this we set up and solve a system of twelve equations in twelve unknowns. Let $a, b, \ldots h, k \ldots n$ represent amounts of chemicals as indicated on the following diagram (fig. 11.3):

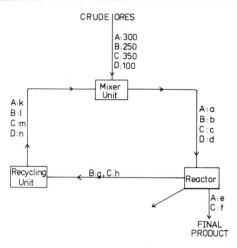

Fig. 11.3   Flow rates in chemical plant.

The twelve equations are:

$$a = 300 + k; \quad b = 250 + l; \quad c = 350 + m; \quad d = 100 + n;$$
$$e = a; \quad f = \tfrac{2}{3}e; \quad g = b; \quad h + f = c;$$
$$k = 0.3(g+h); \quad l = 0.4(g+h); \quad m = 0.2(g+h); \quad n = 0.1(g+h)$$

These can be considerably reduced, to:

$$a = \tfrac{1}{4}(1000 + b + c)$$
$$b = \tfrac{1}{9}(3750 - 4a + 6c)$$
$$c = \tfrac{1}{12}(5250 - 2a + 3b)$$
$$d = \tfrac{1}{15}(1500 - a + 1.5b + 1.5c)$$

Solve these four equations by Gauss–Seidel iteration assuming the chemical plant is starting up with no recycled material initially (so $a_0 = 300$, $b_0 = 250$, $c_0 = 350$, $d_0 = 100$) and so find the rate of production which will build up ($e + f$).

Exercise 11.4   The vibrational energy levels, $G(n)$ of a diatomic molecule such as HCl, are given by

$$G(n) = \omega_e(n + \tfrac{1}{2}) - \omega_e x(n + \tfrac{1}{2})^2 + \omega_e y(n + \tfrac{1}{2})^3 + \omega_e z(n + \tfrac{1}{2})^4 + \ldots$$

where n, the vibrational quantum number, can only take the integer values

$$0, 1, 2, 3, \ldots$$

ω_e is called the fundamental frequency of vibration of the molecule $\omega_e x$, $\omega_e y$, $\omega_e z$, ... are called anharmonicity constants.

Jumps (or transitions) in energy levels can occur, e.g. from $G(0)$ to $G(1)$, when the molecule absorbs infrared radiation. Fig. 11.4 shows some possible jumps between energy levels.

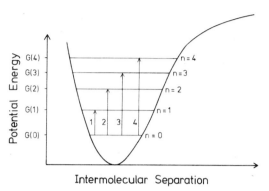

Fig. 11.4   Quantum jumps in HCl.

The transitions labelled 1, 2, 3, 4 are detected as absorption bands in the infrared spectrum of the molecule. For HCl these are found experimentally to be at:

1	$2885{\cdot}981$ cm^{-1}	for	$G(0) \rightarrow G(1)$
2	$5667{\cdot}991$ cm^{-1}	for	$G(0) \rightarrow G(2)$
3	$8346{\cdot}793$ cm^{-1}	for	$G(0) \rightarrow G(3)$
4	$10922{\cdot}86$ cm^{-1}	for	$G(0) \rightarrow G(4)$

We therefore can write

$$\text{I} \quad 2885{\cdot}981 = G(1) - G(0)$$
$$\text{II} \quad 5667{\cdot}991 = G(2) - G(0)$$
$$\text{III} \quad 8346{\cdot}793 = G(3) - G(0)$$
$$\text{IV} \quad 10922{\cdot}86 = G(4) - G(0)$$

Now by substituting for n in the equation for $G(n)$ a set of four equations in the four unknowns ω_e, $\omega_e x$, $\omega_e y$, $\omega_e z$ can be formed.

For example

$$G(0) = \omega_e(\tfrac{1}{2}) - \omega_e x(\tfrac{1}{2})^2 + \omega_e y(\tfrac{1}{2})^3 + \omega_e z(\tfrac{1}{2})^4$$
$$\therefore \quad G(0) = 0{\cdot}5\omega_e - 0{\cdot}25\omega_e x + 0{\cdot}125\omega_e y + 0{\cdot}0625\omega_e z$$

Similarly

$$G(1) = 1{\cdot}5\omega_e - 2{\cdot}25\omega_e x + 3{\cdot}375\omega_e y + 5{\cdot}0625\omega_e z$$

$\therefore$ I becomes

$$2885{\cdot}981 = 1{\cdot}0\omega_e - 2{\cdot}0\omega_e x + 3{\cdot}25\omega_e y + 5{\cdot}0\omega_e z$$

Find similar equations for II, III, IV and then solve the set of four equations to determine ω_e, $\omega_e x$, $\omega_e y$, $\omega_e z$ as accurately as possible.

12. Population dynamics of biological systems

INTRODUCTION

Here we return to a discussion of certain recurrence relations (similar to that of the Fibonacci Rabbit problem discussed in Chapter 4) which are also known as Difference Equations. We will assume initially that a species lives in an enclosed environment, so that there are no migrations, and that there are distinct time units (e.g. seasons) which enable us to deal with each time unit separately, one after the other. If the *birth rate* per unit time is B and the corresponding *death rate* is D then the *growth rate* per unit time will be $B-D$. (A birth rate of 20% would be $B=0.2$ and mean that for every 100 of the population 20 young would be born.) If the population at the beginning of the kth period of time is N_k then the population at the end of the kth period will be:

'Those alive at start of period' + 'Those born in period' − 'Those dying in period', i.e.

$$N_k + BN_k - DN_k$$

This we will take to be the population at the beginning of the next period (the $k+1^{th}$ period)

$$\therefore \quad N_{k+1} = N_k(1+B-D)$$

Recall that we are treating time as separate units (e.g. weeks, months or seasons) rather than as continuous. This is often appropriate for (discrete) biological systems when the populations studied consist of a finite number of individuals and is equivalent to the study of differential equations which correspond to continuous variable processes.

SOME SIMPLE MODELS

BASIC MODEL

In the simplest case B and D are *constant*. It is very easy to calculate the population figures.

Example:

80 Bactrian camels are released in the Australian bush, having been previously imported and used as a means of transport. For the camels $B=0.3$ per season and $D=0.18$ per season (in other words 30% birth rate, 18% death rate). What will the camel herd population be for the first 12 seasons?

In this case $N_{k+1}=N_k(1+0.3-0.18)$, i.e.

$$N_{k+1}=1.12N_k; \quad \text{and} \quad N_1=80$$

Algebraic key strokes	*Display*	*Comment*	*Reverse Polish key strokes*
1·12 [STO]	1·12	Growth factor	1·12 [enter] [enter] [enter]
[×] 80 [×]	89·6	N_2	80 [×]
[RCL] [×]	100·352	N_3	[×]
[RCL] [×]	112·39424	N_4	[×]
[RCL] [×]	125·88154	N_5	[×]
etc.	. . . .		etc.

If all the calculations are performed to pocket calculator accuracy first and then finally rounded off to the nearest integer the set of population figures in Table 12.1 results.

TABLE 12.1

	Population each year	
Year	At beginning	At end
1	80	90
2	90	100
3	100	112
4	112	126
5	126	141
6	141	158
7	158	177
8	177	198
9	198	222
10	222	248
11	248	278
12	278	312

This is the compound interest formula with the exponential type growth curve (fig. 12.1).

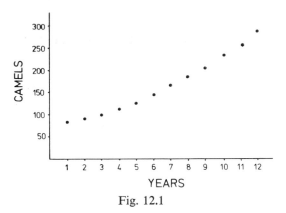

Fig. 12.1

Clearly this is an inadequate long-term mathematical model. No population can increase indefinitely!

LIMITATION MODEL

We will suppose that there is some limit to the food supply or territory so that for a very large population, N_k, the death rate will increase as starvation or overcrowding occurs. One way to introduce this limiting factor is to say that the death rate D is not constant but is in fact proportional to the population size, N_k—the bigger N_k, the bigger the death rate. So we replace constant D by dN_k where d is some constant death rate factor. Our model now becomes

$$N_{k+1} = N_k + BN_k - dN_k \times N_k$$

i.e.

$$N_{k+1} = (1 + B - dN_k)N_k$$

Let us take $N_1 = 80$, $B = 0.3$, $d = 0.0018$ which gives exactly the same starting size and a greater initial growth rate than our previous Basic Model example of the Camels:

$$N_{k+1} = (1.3 - 0.0018N_k)N_k; \quad N_1 = 80$$

Key-stroke details are given on page 203. Table 12.2 on page 204 shows the results obtained.

Although the camel population in the Limitation Model is initially growing faster than in the Basic Model it slows down as 'food shortage' (say) takes its effect and there seems to be a limit to the size of population which the environment can support.

Some things to do

Exercise 12.1   What will happen in the Basic Model if the birth rate is less than the death rate? Test by computing the population levels for

	Algebraic key strokes	Display	Comment	Reverse Polish key strokes	
	80 STO	80.	N_1	80 enter enter	
cycle 1	+/− × ·0018 + 1·3 ×	1·156	growth factor for year 1	·0018 +/− × 1·3 +	cycle 1
	RCL = STO	92·48	N_2	× enter enter	
cycle 2	+/− × ·0018 + 1·3 ×	1·133536	growth factor for year 2	·0018 +/− × 1·3 +	cycle 2
	RCL = STO	104·8294	N_3	× enter enter	
cycle 3	+/− × ·0018 + 1·3 ×	1·111307	growth factor for year 3	·0018 +/− × 1·3 +	cycle 3
	RCL = STO	116·49766	N_4	× enter enter	
	etc.	· · · ·	· · · · ·	etc.	

(*Note*: Memories could be used to hold the coefficients 1·3 and − 0·0018 if available and considered worthwhile.)
The results, together with those for the Basic Model, are shown in Table 12.2.

10 years, assuming a birth rate of 0·31, and a death rate of 0·49 for an initial population of 100. Sketch the resulting curve through the points. What kind of curve is it?

TABLE 12.2

Beginning of year	Limitation model	Basic model
1	80	80
2	92	90
3	105	100
4	116	112
5	127	126
6	136	141
7	144	158
8	150	177
9	154	198
10	158	222
11	160	248
12	162	278
13	163	312
⋮	⋮	⋮
∞	?	∞

Exercise 12.2   In the Limitation Model example the population appears to be tending to a limit. What do you think it is?

(a) One method to find the answer is to keep computing more 'cycles'—see what has happened by another five years or so.

(b) An analytic method is to argue that if the population is going to level off then the growth factor $(1 + B - dN_k)$ must be 1. In this case we need $(1·3 - 0·0018N_k) = 1$ so what is N_k at the limit?

Exercise 12.3   A ship carrying some white mice in crates is shipwrecked on an island, and 1000 mice survive. The birth rate per month is 0·6 (i.e. 60 mice born per 100 mice) and the death rate, D, is initially 0·95 per month. Predict the future monthly white mouse population

(a) Using the Basic Model

(b) Using the Limitation Model, remembering that if $D = 0·95$ then

$$d = \frac{0·95}{1000} = 0·00095.$$

Exercise 12.4   To combat a plague of aphids, 20,000 ladybirds are artificially introduced into an area.

Assuming that the birth rate, B, is 1·9 per time unit and the death rate factor for the Limitation Model, d, is 0·000012:

(a) Plot a graph of the ladybird population for the next 8 time units

(b) What happens if the birth rate is 3·1 instead?

Exercise 12.5

(a) It should be clear from the preceding examples that the behaviour of the population size in the Limitation Model can:

   (i) increase to an equilibrium level [camel example in Table 12.2]

   (ii) decrease to an equilibrium level [Exercise 12.3]

   (iii) have decreasing oscillations about an equilibrium level [Exercise 12.4 (a)]

   (iv) have increasing oscillations until extinction occurs [Exercise 12.4 (b)]

   (v) One further category—steady oscillations—is exemplified by the following example [Exercise 12.5 (b)].

(b) In an experiment 120 drosophila flies are introduced into a population cage and a limited food supply maintained. The birth rate is known to be 2·4 per fortnight and the death rate, D, is initially 1·44 per fortnight (i.e. $d = 1·44/120 = 0·012$). Find the population sizes each fortnight for the first 24 weeks of the experiment.

SOME FURTHER MODELS

It may well be the case that the current food supply is partially determined by how much food was consumed some time ago (by the then population). For example if in one year a group of herbivores eat up all the grass in their territory there is likely to be a poor crop of grass the following season with serious effect on the herbivores.

LAST SEASON MODEL

Let the population of herbivores in a year be N_k then if the (constant) birth rate is B, and the death rate D is proportional to last year's population size (i.e. higher if there was a large population last year) then we get, writing dN_{k-1} for D:

$$N_{k+1} = N_k + BN_k - dN_{k-1} \times N_k$$

i.e.

$$N_{k+1} = (1 + B - dN_{k-1})N_k$$

Computation is again fairly straightforward but the *two* previous years' population levels (N_k and N_{k-1}) are both needed to determine the next year's population, N_{k+1}, so two memories or re-entry of data are required to proceed from year to year if an algebraic machine is used. Judicious use of the stack avoids these problems for Reverse Polish machines.

Example: Sheep Population

Suppose that as a zoological experiment a flock of 100 sheep is introduced one autumn to a small uninhabited island, and that by the next autumn the flock is observed to have grown to 130. Thereafter the flock is left entirely alone.

(a) If the birth rate, B, is 0·6 and the death rate factor, d, is 0·002 what will happen to the flock over the next decade? Is there a limit to the size of the flock?

The basic calculation necessary is, given $N_1 = 100$, $N_2 = 130$, to compute

$$N_{k+1} = (1·6 - 0·002N_{k-1})N_k \quad \text{for } k = 2, 3, \ldots$$

Algebraic key strokes	Display	Comment
130 [STO]	130·	N_2
cycle 1 { ·002 [+/−] [×] 100		
[+] 1·6 [×]	1·4	growth rate for year
[RCL] [=] [STO]	182·	N_3
cycle 2 { ·002 [+/−] [×] 130		
[+] 1·6 [×]	1·34	growth rate for year
[RCL] [=] [STO]	243·88	N_4
etc.	. . .	. . .

Reverse Polish key strokes	Display	Comment
130 [enter] [enter]	130·	N_2
100 [enter]	100	N_1
cycle 1 { ·002 [+/−] [×] 1·6		
[+]	1·4	growth rate for year
[×] [enter] [enter]	182·	N_3
[R↓] [R↓] [R↓]	130	N_2 recovered (see note)
cycle 2 { ·002 [+/−] [×] 1·6		
[+]	1·34	growth rate for year
[×] [enter] [enter]	243·88	N_4
[R↓] [R↓] [R↓]	182·	N_3 recovered (see note)

Reverse Polish Note: On some machines (e.g. Corvus 500) only *two* [R↓] key strokes should be used here, and the display is therefore different.

The population levels are shown in Table 12.3.

TABLE 12.3

Year	1	2	3	4	5	6	7	8	9	10	11	12	13
Size	100	130	182	244	301	335	334	311	289	283	289	299	306

And the graph, as shown in fig. 12.2, levels out at 300.

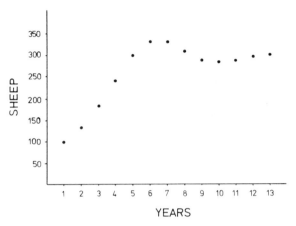

Fig. 12.2   Fluctuations in sheep population derived from Last-Season Model.

(b) What would happen if 500 (or 1000) immature sheep—i.e. not breeding until one season later—had been introduced?

For the case of 500 sheep $N_1 = 500$, $N_2 = 500$ and then despite breeding starting the shortage of food would reduce the population:

$$N_3 = (1 \cdot 6 - 0 \cdot 002 \times 500) \times 500 = 300$$
$$N_4 = (1 \cdot 6 - 0 \cdot 002 \times 500) \times 300 = 180$$
$$N_5 = (1 \cdot 6 - 0 \cdot 002 \times 300) \times 180 = 180$$

and then the population begins to rise again (for the reader to investigate).

For the case of 1000 sheep $N_1 = 1000$, $N_2 = 1000$

$$N_3 = (1 \cdot 6 - 0 \cdot 002 \times 1000) \times 1000 = -400$$

so the flock would destroy all the pasture and die out through starvation.

MULTI-SEASON MODEL

The Last Season Model takes account of the previous season only but not the current one, or earlier ones, in the death rate. Moving a step towards a more realistic multi-season model—which would be one taking account of several past seasons which will have influenced the age structure of the population—we shall re-introduce the present population as a factor in the death rate and also the population size of the last season but one. The difference equation now has three seasons' population sizes affecting the next season's population size:

$$N_{k+1} = N_k + BN_k - (dN_k + eN_{k-1} + fN_{k-2})N_k$$

where B is the birth rate and d, e, f are death rate factors.

For example, with $N_1 = 300$, $N_2 = 340$, $N_3 = 310$, $B = 0·8$, $d = 0·0015$, $e = 0·0010$, $f = 0·0005$ the population curve is as in fig. 12.3.

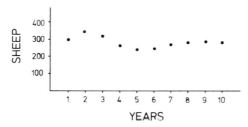

Fig. 12.3   Fluctuations in sheep population derived from Multi-Season Model.

Some things to do

Exercise 12.6   Use the Last Season Model on the island sheep problem (page 205) to determine the largest initial population size that does not lead to extinction of the flock:

(*a*) if there is no breeding the first year

(*b*) if breeding occurs in the first year.

Exercise 12.7   For the Multi-Season Model (page 207) with $N_1 = 300$, $N_2 = 340$, $N_3 = 310$, $B = 0·8$, what effect does it have to alter the individual values of d, e, f keeping their sum constant?
    Try the following:

	d	e	f
(*a*)	0·003	0	0
(*b*)	0	0·003	0
(*c*)	0	0	0·003
(*d*)	0·001	0·001	0·001

Exercise 12.8   A variant on the Fibonacci Model is to assume that the *birth rate* declines as the population grows. The birth rate term in the Fibonacci Model $N_{k+1} = N_k + N_{k-1}$ is the N_{k-1} term (really $1 \times N_{k-1}$)

so the 1 might be replaced by $e^{-0\cdot1N_k}$ which for small N_k is ≈ 1 and for large N_k is ≈ 0

$$\therefore \quad N_{k+1} = N_k + e^{-0\cdot1N_k} \times N_{k-1}$$

Taking $N_1 = 1$, $N_2 = 2$ compute population sizes rounding off at each stage, and comparing with the Fibonacci sequence. Is there a limit?

Exercise 12.9   Another variant on the Fibonacci Model would include deaths of some of the younger rabbits (say 30%). Then we might have

$$N_{k+1} = 0\cdot7N_k + e^{-0\cdot02N_k} \times N_{k-1}$$

Show that this leads to an equilibrium population of size 59 to 61 whatever N_1 and N_2 are taken to be (round off at each stage).

COMPETING SPECIES

A more interesting situation arises when two species interact—predator and prey; host and parasite; symbiotic partners; competitors.

In such cases a *pair* of difference equations, one for each species, is needed.

We will look at the foxes and rabbits situation. Assuming an unlimited supply of food and no deaths the rabbits would multiply exponentially, and their difference equation, where R_k means the number of rabbits at period k, would be for birth rate b:

$$R_{k+1} = R_k + bR_k$$

However, there are foxes about and it can be assumed that a proportion of rabbit–fox encounters will result in the death of a rabbit. The number of such deaths will be more if there are more rabbits (easier to come by) and more if there are more foxes (more hunters) and the simplest assumption is that deaths will be proportional to each of these numbers. So the rabbit deaths will be $d \times R_k \times F_k$, where F_k is the number of foxes about at period k and d is some (positive) constant.

$$\therefore \quad R_{k+1} = R_k + bR_k - dR_kF_k$$

Now for the foxes the position is more or less reversed. What is bad for rabbits is good for foxes! So the foxes will prosper the more that rabbits die in encounters. This suggests that the foxes will prosper according to $F_{k+1} = aR_kF_k$ for some (positive) constant a.

The more rabbits there are, the more successfully will the foxes reproduce, and the more foxes there are, the more offspring they will rear. Without rabbits the foxes, we assume, will die out and this is seen to be so because if $R_k = 0$ then $F_{k+1} = 0$ as required. We thus have the pair of difference equations:

$$R_{k+1} = (1+b)R_k - dR_kF_k$$
$$F_{k+1} = aR_kF_k$$

Taking as an example the case of 8 foxes and 500 rabbits where the birth rate for rabbits, b, is $0 \cdot 1$ and the death rate constant, d, is $0 \cdot 01$ and the growth rate constant for foxes, a, is $0 \cdot 003125$ we have

$$R_{k+1} = 1 \cdot 1 R_k - 0 \cdot 01 R_k F_k$$
$$F_{k+1} = 0 \cdot 003125 R_k F_k$$
$$R_1 = 500$$
$$F_1 = 8.$$

The calculator can be used most effectively here to perform the calculations. One equation is used, then the other, and so on alternately.

The following Algebraic logic procedure is tempting if two memories are available:

	Algebraic key strokes	Display	Comment
	(a) 500 $\boxed{\text{STO 1}}$	500·	R_1 stored in Memory 1
	(b) 8 $\boxed{\text{STO 2}}$	8·	F_1 stored in Memory 2
First Equation	(c) ·01 $\boxed{+/-}$ $\boxed{\times}$ $\boxed{\text{RCL 2}}$	8·	$-0 \cdot 01 F_1$
	(d) $\boxed{+}$ 1·1 $\boxed{=}$	1·02	$1 \cdot 1 - 0 \cdot 01 F_1$
	(e) $\boxed{\times}$ $\boxed{\text{RCL 1}}$ $\boxed{=}$ $\boxed{\text{STO 1}}$	510·	R_2
Second Equation	(f) $\boxed{\times}$ ·003125	·003125	$0 \cdot 003125 R_2$
	(g) $\boxed{\times}$ $\boxed{\text{RCL 2}}$ $\boxed{=}$ $\boxed{\text{STO 2}}$	12·75	F_2 (rounds to 13)

Now there is in fact a mistake here (can you spot it?). In calculating line (f) we should be using the *old* rabbit population size (500) not the newly calculated and stored value (510). It *is* possible to perform the calculations without re-entering the population sizes provided one memory is available but it is probably not worth the effort. It is better to keep matters simple by writing down the population levels as they are calculated (rounding off) and then use these values as needed. (Memories might be better used for storing the constants in the equations.)

	Algebraic key strokes	Display	Comment
Equation 1	·01 $\boxed{+/-}$ $\boxed{\times}$ 8		
	$\boxed{+}$ 1·1 $\boxed{\times}$ 500		
	$\boxed{=}$	510	R_2

(*continued*)

	Algebraic key strokes	Display	Comment
Equation 2	·003125 $\boxed{\times}$ 500		
	$\boxed{\times}$ 8 $\boxed{=}$	12·5	F_2 (13 when rounded)
Equation 1	·01 $\boxed{+/-}$ $\boxed{\times}$ 13		
	$\boxed{+}$ 1·1 $\boxed{\times}$ 510		
	$\boxed{=}$	494·7	R_3 (495 when rounded)
Equation 2	·003125 $\boxed{\times}$ 510		
	$\boxed{\times}$ 13 $\boxed{=}$	20·7	F_3 (21 when rounded)
	etc.	. . . . . .	

Memories can of course be used to hold F_k and R_k and it is particularly helpful to calculate the Fox equation first and store $R_k F_k$ for use in the Rabbit equation.

Reverse Polish procedure is rather different and the whole process can be achieved with no re-entries of F_k and R_k values, using just one memory. This actually does not use the rounded off values, but the differences are slight. Rounded off values could of course be re-entered if preferred.

	Reverse Polish key strokes	Display	Comment
	8 $\boxed{STO}$	8·	F_1
	500 $\boxed{enter}$ $\boxed{enter}$ $\boxed{enter}$	500·	R_1
	$\boxed{RCL}$ $\boxed{\times}$ $\boxed{enter}$ $\boxed{enter}$	4000·	$R_1 F_1$
	·003125 $\boxed{\times}$ $\boxed{STO}$	12·50	F_2 (now stored)
	$\boxed{R\downarrow}$	4000·	$R_1 F_1$ (recovered)
Cycle	·01 $\boxed{+/-}$ $\boxed{\times}$	−40·	$-0·01 R_1 F_1$
	$\boxed{R\downarrow}$ $\boxed{R\downarrow}$ $\boxed{R\downarrow}$	500·	R_1 (recovered)
	1·1 $\boxed{\times}$	550·	$1·1 R_1$
	$\boxed{+}$ $\boxed{enter}$ $\boxed{enter}$ $\boxed{enter}$	510·	R_2

and then the cycle is repeated.

Table 12·4 gives the results (i) when rounded off F_k and R_k values are used in the computational procedures ('Continuous Rounding'), and (ii) when unrounded values are used, with rounding only performed when recording the computed values ('Final Rounding').

So in this case (Table 12.4) the foxes increase at first but then kill off so many rabbits that they starve to death . . . and the rabbit population survives!

TABLE 12.4

k	(i) Continuous Rounding		(ii) Final Rounding	
	R_k	F_k	R_k	F_k
1	500	8	500	8
2	510	13	510	13
3	495	21	497	20
4	441	32	448	31
5	344	44	354	43
6	227	47	236	48
7	143	33	146	35
8	110	15	110	16
9	105	5	103	5
10	110	2	108	2
11	119	1	117	1
12	130	0	127	0

Changing the parameters b, d, a, R_1, F_1, generally leads to the extermination of one or other species and the model is 'unstable'. Unless an equilibrium situation is established initially (e.g. 320 rabbits, 10 foxes for the parameters as given) the fluctuations get worse and worse.

An adjustment to the model is to assume that the foxes have a natural death rate, so a term $-c \times F_k$ could be added to give:

$$R_{k+1} = (1+b)R_k - dR_kF_k$$
$$F_{k+1} = aR_kF_k - cF_k$$

For $b=0\cdot1$, $d=0\cdot01$, $a=0\cdot002$, $c=0\cdot03$ we have

$$R_{k+1} = 1\cdot1R_k - 0\cdot01R_kF_k$$
$$F_{k+1} = 0\cdot002R_kF_k - 0\cdot03F_k$$

With 515 rabbits and 10 foxes equilibrium occurs but for other values the tendency is for the foxes to increase too fast for their own good.

Some things to do

Exercise 12.10   For the model

$$R_{k+1} = 1\cdot2R_k - 0\cdot01R_kF_k$$
$$F_{k+1} = 0\cdot002R_kF_k$$

there are initially 10 foxes.

Show that if there are 1000 rabbits, then the rabbits all perish first but that with 500 rabbits the foxes die out although the rabbits remain quite plentiful.

Exercise 12.11   For the model described earlier in the text:

$$R_{k+1} = 1 \cdot 1 R_k - 0 \cdot 01 R_k F_k$$
$$F_{k+1} = 0 \cdot 002 R_k F_k - 0 \cdot 03 \, F_k$$

show that the fox population remains fairly constant if there are 500 rabbits and 10 foxes initially but that it is really unstable.

Exercise 12.12   The instability of the simpler models may be due to the unrealistic assumption that the rabbits cannot themselves run short of food. To take this into account another factor should be included to represent competition for the food: $-c \times R_k \times R_k$. Thus we have the pair of difference equations

$$R_{k+1} = (1 + b)R_k - d R_k F_k - c R_k R_k$$
$$F_{k+1} = a R_k F_k$$

This added factor tends to limit the growth of the rabbit population and so avoids the over-increase in the fox population which leads to disaster. Show that if $b = 0 \cdot 1$, $d = 0 \cdot 01$, $c = 0 \cdot 0001$, $a = 0 \cdot 002$, 500 rabbits and 10 foxes will lead to a balance but that if $b = 0 \cdot 25$ disaster ensues.

Exercise 12.13   Another predator–prey model is based on the assumption that predators too will be competing with each other—for territory, mates, food, shelter—and so their death rate should not be constant but proportional to their numbers. This leads to the model

$$R_{k+1} = (1 + b)R_k - d R_k F_k - c R_k R_k$$
$$F_{k+1} = a R_k F_k - e F_k F_k$$

This is a much more stable model. Show that for $b = 0 \cdot 25$, $d = 0 \cdot 01$, $c = 0 \cdot 0001$, $a = 0 \cdot 002$, $e = 0 \cdot 02$ with 500 rabbits and any reasonable initial number of foxes the model leads to equilibrium.

Exercise 12.14   The difference equations for a host–parasite interaction —such as a fly which lays its eggs in the caterpillars of a certain species of moth—have been shown to be

$$H_{k+1} = a H_k e^{-b P_k}$$
$$P_{k+1} = a H_k (1 - e^{-b P_k})$$

Taking the case where $a = 1 \cdot 2$, $b = 0 \cdot 0002$ when initial density of hosts is 10000 per hectare and of parasites is 1000 per hectare show that large oscillations in the population occur. This has been observed in reality by Varley and is discussed by J. Maynard Smith (1968).

Exercise 12.15   *An open-ended problem: Snowy Owls and Lemmings*
The predator–prey relationship between snowy owls and lemmings in the Arctic is a simple case because the lemmings form a large part of the owls' diet and there is little alternative food supply. The following approximate figures have been suggested (you may alter these):

(i) A mature female lemming (1 year old) may produce about four litters per year of about four young each.

(ii) A mature female snowy owl (2 years old) will lay about four eggs in a season.

(iii) The average life-span for a lemming is about two years.

It might be surmised that:

(iv) When lemmings are plentiful two of the four owl eggs are hatched and reared to mature birds (taking two years) but when food is very scarce no youngsters survive.

(v) The death rate for mature owls is 30% per annum.

(vi) A mature owl needs 800 lemmings per year and an immature bird needs on average 400 lemmings per year.

Can you set up some possible difference equations to describe this system—making any necessary assumptions and modifications—which will correspond to the real-life situation in which population fluctuations occur at about 5 yearly periods?

13. A numerical approach to calculus

NUMERICAL DIFFERENTATION

INTRODUCTION

From a graphical point of view differentiation is to do with the *gradient* of some curve. We all know that for a *straight line* the gradient is constant—the slope stays the same all along the line. How can we find the gradient at a point on *any* curve? For example, what is the gradient of $y=x^3$ at $x=1\cdot7$ in fig. 13.1?

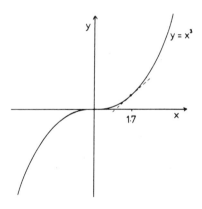

Fig. 13.1

One approach is to concentrate on the curve *very near to* $x=1\cdot7$, finding two points, P, Q, on the curve in the neighbourhood of $1\cdot7$ and calculating the gradient of the chord PQ (see fig. 13.2) which should give an estimate for that of the curve actually *at* $1\cdot7$ since $y=x^3$ is a smooth continuous curve (as are all polynomial functions).

The formula for the gradient of the straight line joining $P(x_P, y_P)$ and $Q(x_Q, y_Q)$ is given by

$$\frac{y_Q - y_P}{x_Q - x_P}$$

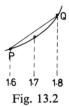

Fig. 13.2

If we take P to be at $x=1\cdot6$ and Q to be at $x=1\cdot8$ then we can find the y co-ordinates of P and Q from the equation $y=x^3$ and then use these values to calculate the gradient estimate:

Algebraic key strokes	Display	Comment
$1\cdot6$ $\boxed{y^x}$ 3 $\boxed{=}$ $\boxed{\text{STO}}$	$4\cdot096$	y_P
$1\cdot8$ $\boxed{y^x}$ 3 $\boxed{=}$	$5\cdot832$	y_Q
$\boxed{-}$ $\boxed{\text{RCL}}$ $\boxed{=}$	$1\cdot736$	y_Q-y_P
$\boxed{\div}$ $\cdot2$ $\boxed{=}$	$8\cdot68$	$(y_Q-y_P)/(x_Q-x_P)$

Reverse Polish key strokes	Display	Comment
$1\cdot8$ $\boxed{\text{enter}}$ 3 $\boxed{y^x}$	$5\cdot832$	y_Q
$1\cdot6$ $\boxed{\text{enter}}$ 3 $\boxed{y^x}$	$4\cdot096$	y_P
$\boxed{-}$	$1\cdot736$	y_Q-y_P
$\cdot2$ $\boxed{\div}$	$8\cdot68$	$(y_Q-y_P)/(x_Q-x_P)$

Now we might try $x_P=1\cdot69$ and $x_Q=1\cdot71$ to 'close in' on the point $x=1\cdot7$, and so on. . . . Table 13.1 shows the results of such calculations (the reader may like to check them).

TABLE 13.1

	Point P		Point Q		Gradient estimate
h	x_P	y_P	x_Q	y_Q	$(y_Q-y_P)/(x_Q-x_P)$
$0\cdot1$	$1\cdot6$	$4\cdot096$	$1\cdot8$	$5\cdot832$	$8\cdot68$
$0\cdot01$	$1\cdot69$	$4\cdot826809$	$1\cdot71$	$5\cdot000211$	$8\cdot6701$
$0\cdot001$	$1\cdot699$	$4\cdot9043351$	$1\cdot701$	$4\cdot9216751$	$8\cdot670001$
$0\cdot0001$	$1\cdot6999$	$4\cdot9121331$	$1\cdot7001$	$4\cdot9138671$	$8\cdot6700000$

The sequence of estimates has clearly converged to the correct value $8\cdot67$. (Those who can differentiate analytically can verify this!)

There is no necessity to pick P and Q at *equal* x distances from $1\cdot7$. Indeed one of them can be $1\cdot7$ itself although this is in fact rather less accurate.

Our computing method above can be written as:

$$\text{`gradient of } x^3 \text{ at } x=a\text{'} \approx \frac{(a+h)^3-(a-h)^3}{2h}$$

In our example $a=1\cdot7$ and h was initially $0\cdot1$, then $0\cdot01$, then $0\cdot001$, and then $0\cdot0001$.

More generally the function will not be $y=x^3$ but $y=f(x)$. In that case we have:

$$\text{`gradient of } f(x) \text{ at } x=a\text{'} \approx \frac{f(a+h)-f(a-h)}{2h}$$

Example 1

Estimate the gradient of $y=\sin x+x^2$ at $x=a=0\cdot5$ taking $h=0\cdot01$. Here $f(x)=\sin x+x^2$ and $a+h=0\cdot51$, $a-h=0\cdot49$ and we must calculate

$$\frac{\{\sin 0\cdot51+(0\cdot51)^2\}-\{\sin 0\cdot49+(0\cdot49)^2\}}{0\cdot02}$$

Algebraic key strokes	Display	Comment	Reverse Polish key strokes
Select radian mode			Select radian mode
$\cdot51$ $\boxed{\sin}$ $\boxed{+}$ $\cdot51$			$\cdot51$ $\boxed{\text{enter}}$ $\boxed{\sin}$
$\boxed{x^2}$ $\boxed{-}$	$\cdot74827725$	$f(0\cdot51)$	$\boxed{x\rightleftharpoons y}$ $\boxed{x^2}$ $\boxed{+}$
$\cdot49$ $\boxed{\sin}$ $\boxed{-}$ $\cdot49$			$\cdot49$ $\boxed{\text{enter}}$ $\boxed{\sin}$
$\boxed{x^2}$		$f(0\cdot51)-$	$\boxed{x\rightleftharpoons y}$ $\boxed{x^2}$ $\boxed{+}$
$\boxed{=}$	$\cdot03755136$	$f(0\cdot49)$	$\boxed{-}$
$\boxed{\div}$ $\cdot02$ $\boxed{=}$	$1\cdot8775679$	gradient	$\cdot02$ $\boxed{\div}$

The analytic answer is $\cos x+2x$ where $x=0\cdot5$, and the reader can evaluate this to get an indication of the accuracy of the numerical method.

Example 2

Estimate the gradient of $\cos 50x$ at $x=1\cdot3$ using $h=0\cdot02$. Using a calculator produces $-34\cdot787617$.

Now the analytic answer is $-50\sin 50x$ where $x=1\cdot3$. Evaluating this directly gives $-41\cdot341434$ and we have a disaster on our hands! Why is the numerical method so inaccurate? The function $\cos 50x$ oscillates very rapidly and so its gradient is changing very rapidly too. Taking $h=0\cdot02$ is too large for this particular function. Taking $h=0\cdot002$ gives a far more satisfactory value $(-41\cdot27\ldots)$.

DERIVING A FORMULA

Differential calculus seeks to answer the question 'what happens as h tends to 0?' for the *general* point $x = a$ rather than for $x = 1.7$ or some other particular numerical value. In the language of Analysis the value of

$$\text{Lim}_{h \to 0} \left\{ \frac{f(a+h) - f(a-h)}{2h} \right\}$$

is sought for general a.

The difficulty with numerical differentiation is that the methods inevitably require one to subtract two very similar numbers and divide by a small number, both of which processes tend to magnify inaccuracies and lead to loss of significant figures. If h is taken too small then results can be worse than using a larger value. In difficult cases (e.g. $\sin 100x$) a compromise has to be reached, and only a limited accuracy is possible if using a pocket calculator or computer.

The last point is well illustrated for the example already discussed: i.e. the gradient of $y = x^3$ at $x = 1.7$. If we use the very small $h = 0.0000001$ then we must calculate

$$\frac{1.7000001^3 - 1.6999999^3}{0.0000002}$$

which, on a simple 8-digit calculator may give 8.5—a much less accurate result than using $h = 0.1$!

Some things to do

Use the formula $\dfrac{f(a+h) - f(a-h)}{2h}$ to estimate the gradients of the following:

Exercise 13.1 $f(x) = \sqrt{x}$ at $x = 5.6$ taking $h = 0.01$.

Exercise 13.2 $f(x) = \tan x$ at $x = 1.57$, taking various h.

Exercise 13.3 $f(x) = \ln [\cos (\sin \{x^2\})]$ at $x = 0.724$, taking $h = 0.0005$.

Exercise 13.4 $f(x) = \sin (1/x)$ at $x = 0.01$ taking various h.

Exercise 13.5 $y = x^{10}$ at $x = 11.253$ taking various h to maximize accuracy.

A SIMPLE FORMULA

If the pair of points P, Q used in estimation of the gradient are chosen so that P is actually at the point of interest then the formula $(y_Q - y_P)/(x_Q - x_P)$ becomes

$$\frac{f(a+h) - f(a)}{h}$$

as illustrated in fig. 13.3.

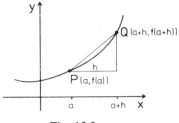

Fig. 13.3

Example:

Use $[f(a+h)-f(a)]/h$ to estimate the gradient of $y=2^x$, at $x=1\cdot5$ taking $h=0\cdot01$.

We require

$$\frac{2^{1\cdot51}-2^{1\cdot5}}{0\cdot01}$$

and the calculator yields $1\cdot9673266$.

For comparison the formula $[f(a+h)-f(a-h)]/2h$ using the same h gives $1\cdot9605320$ and the correct value is $1\cdot9605163$ to 8 d.p.

A MORE ACCURATE FORMULA

A rather more involved but generally better formula for estimating derivatives (or gradients of curves) is

$$\frac{-f(a+2h)+8f(a+h)-8f(a-h)+f(a-2h)}{12h}$$

Note that *four* function evaluations are needed so there is twice the work to do. We shall now repeat the problem we did before using $[f(a+h)-f(a-h)]/2h$.

Example:

Estimate the gradient of $f(x)=x^3$ at $x=a=1\cdot7$ using $h=0\cdot1$.

$$a+2h=1\cdot9, \quad a+h=1\cdot8, \quad a-h=1\cdot6, \quad a-2h=1\cdot5$$

therefore the estimate is $[-f(1\cdot9)+8f(1\cdot8)-8f(1\cdot6)+f(1\cdot5)]/1\cdot2$.

Now $f(x)=x^3$ so we must compute

$$[-(1\cdot9)^3+8\times(1\cdot8)^3-8\times(1\cdot6)^3+(1\cdot5)^3]/1\cdot2$$

Evaluating this on a calculator gives $8\cdot67$ exactly! Clearly we have a powerful method here, but it does not always give such a good answer. Because of the simple numbers involved no round off errors have occurred and the method is theoretically exact for polynomials up to fourth degree (i.e. for all linear, quadratic, cubic and quartic polynomials).

Some things to do

Exercise 13.6   Use each of the three formulae to estimate the gradient of $f(x) = e^{x^2}$ at $x = 1 \cdot 15$ taking $h = 0 \cdot 08$ and then $h = 0 \cdot 04$ and then $h = 0 \cdot 02$. Set out your results in tabular form:

	GRADIENT ESTIMATES		
Formula	$h=0 \cdot 08$	$h=0 \cdot 04$	$h=0 \cdot 02$
(i) $\dfrac{f(a+h)-f(a)}{h}$			
(ii) $\dfrac{f(a+h)-f(a-h)}{2h}$			
(iii) $\dfrac{-f(a+2h)+8f(a+h)-8f(a-h)+f(a-2h)}{12h}$			

Exercise 13.7   The errors in these three formulae are approximately proportional to (i) h, (ii) h^2, and (iii) h^4 respectively. This means that if h is halved then the errors should go down by a factor (i) $\frac{1}{2}$, (ii) $(\frac{1}{2})^2$, (iii) $(\frac{1}{2})^4$.

(a) Complete the table below based on the results of Exercise 13.6, given that the correct value of the gradient is $8 \cdot 6314208$ to 8 s.f.

	ERRORS IN ESTIMATES			
Formula	$h=0 \cdot 08$	$h=0 \cdot 04$	$h=0 \cdot 02$	Theoretical Factor
(i)				$\frac{1}{2}$
(ii)				$\frac{1}{4}$
(iii)				$\frac{1}{16}$

(b) Check to see if the errors do go down by the theoretical factors as h is halved.

Exercise 13.8   An interesting property of the function e^x is that the gradient at any point x is itself e^x. For example at $x=0.317$, $e^x = e^{0.317} = 1.3730026$ and the gradient $= 1.3730026$ also, to 8 s.f.

Check this property by using the gradient formula

$$[f(a+h)-f(a-h)]/2h$$

where $h=0.001$, for various values of a.

Exercise 13.9   It is often shown (but perhaps not very convincingly!) in calculus courses that the gradient of the curve $y=f(x)=\sin x$ is $\cos x$ where x is in radians.

For example at $x=0.714$.

$$\text{gradient} \approx \frac{\sin 0.715 - \sin 0.714}{0.001} = 0.7554 \text{ to 4 d.p.}$$

As a check: $\cos 0.714$ by calculator $= 0.7557$ to 4 d.p.

Investigate the following derivatives (i.e. gradients) by numerical techniques:

(a) The derivative of $\cos x$ is $-\sin x$

(b) The derivative of $\tan x$ is $1/\cos^2 x$

(c) The derivative of $\ln x$ is $1/x$

(d) The derivative of $\arctan x$ is $1/(1+x^2)$

(e) The derivative of $\sinh x$ is $\cosh x$.

Exercise 13.10   Find the gradient of $f(x)=\sin[x^2 \ln(\cos\{\tan(\sqrt{x})\})]$ at $x=0.5$, remembering to work in radian mode. (Ask someone to do it using calculus!)

NUMERICAL INTEGRATION (QUADRATURE)

INTRODUCTION

Whereas it is always possible (if somewhat tedious!) to work out the derivative or gradient of any analytic function $f(x)$ in terms of the elementary functions (i.e. those available on a scientific calculator) this is not true for the reverse process of integration.

We can analytically *differentiate* e^{-x^2} and $\sin(\sqrt{x})$ and $(\sin x)/x$ for example, but we cannot analytically *integrate* any of them! However, *numerical* techniques can readily be found to evaluate definite integrals, such as these:

$$\int_{0.1}^{0.3} e^{-x^2}\, dx, \qquad \int_{0}^{0.3} \sin(\sqrt{x})\, dx, \qquad \int_{0}^{15} \frac{\sin x}{x}\, dx$$

What is definite integration anyway? It can usually be considered as finding the area under a curve between certain limits ($x=a$ and $x=b$ say) and bordered by the x-axis. The word 'definite' here simply means that the limits have been defined as particular numbers. For example,

$$\int_1^3 \exp\left(\sqrt{x}\right) dx$$

may be represented as in fig. 13.4.

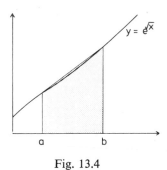

Fig. 13.4

For those unfamiliar with the notation there is no need to despair. Basically it is quite straightforward. The $\int$ symbol is a long S standing for 'Summation' because integration is a process of summing up many small areas. The dx symbol tells us what the variable is: dx (when it is x) or dt (when it is t) and so on. Anything else after the $\int$ symbol is the function—which describes the particular curve, the area under which we require between the limits indicated. So to solve

$$\int_5^{10} x^3 \, dx$$

means 'find the area under the curve $y=x^3$ (or $f(x)=x^3$) starting from $x=5$ and ending at $x=10$', illustrated in fig. 13.5.

This interpretation of integration leads us to some elementary numerical integration formulae, the simplest of which are the 'Midpoint Rule' and the 'Trapezium Rule'.

We shall now take up the problem of evaluating $\int_1^3 \exp\left(\sqrt{x}\right) dx$ which is depicted in fig. 13.4. This is an integral for which no analytic formula exists—it cannot be answered in terms of the elementary functions—and so it must be solved numerically.

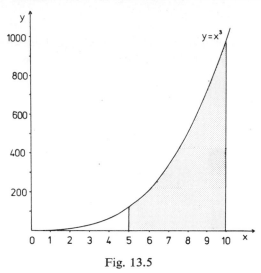

Fig. 13.5

MIDPOINT RULE

We wish to approximate the exact area (shown in fig. 13.4) by a rectangular area which we can very easily evaluate (fig. 13.6). If we take the value of the function midway between the limits ($x=a$ and $x=b$) we shall have some sort of 'average height', $f[(a+b)/2]$,

$$A \approx (b-a) \times f[(a+b)/2] \dots \text{\underline{MIDPOINT RULE}}$$

In our example $b=3$, $a=1$, $f(x)=\exp(\sqrt{x})$ so

$$A \approx (3-1)\exp(\sqrt{((1+3)/2)})$$

$$\therefore \quad A = 8 \cdot 23 \text{ to 2 d.p.}$$

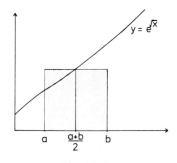

Fig. 13.6

All the information is being derived from one point which may be quite unrepresentative. Perhaps taking two points would be better. . . .

TRAPEZIUM RULE

This time we approximate the area by the trapezium with the parallel sides $f(a)$ and $f(b)$ (fig. 13.7). The appropriate area formula gives

$$A \approx ((b-a)/2) \times (f(a)+f(b)) \ldots \underline{\text{TRAPEZIUM RULE}}$$

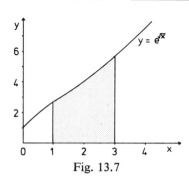

Fig. 13.7

For the example $\int_1^3 \exp(\sqrt{x})\, dx$ we get

$$A \approx ((3-1)/2) \times (\exp(\sqrt{1}) + \exp(\sqrt{3})) = 8.37 \text{ to 2 d.p.}$$

And now we have two rather different results (8·23 and 8·37) and do not know which, if either, is accurate! Clearly more information is needed!

Some things to do

(Remember to work in radians for these problems)

Exercise 13.11   Estimate $\int_{0.5}^{1} \tan x\, dx$ using the two rules.

Exercise 13.12   Estimate $\int_{1}^{1.5} \ln(\sin(x)+x^2)\, dx$ using the two rules.

COMPOUND RULES

One way to improve accuracy is to split up the interval of integration $[a, b]$ into sub-intervals, and to split the area up into a series of strips and apply a rule to each strip.

COMPOUND MIDPOINT RULE

For illustrative purposes we shall use four strips to evaluate the integral $\int_1^3 \exp(\sqrt{x})\, dx$ as shown in fig. 13.8.

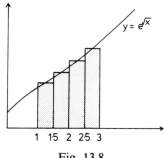

Fig. 13.8

Strip 1   [1, 1·5] $A_1 \approx (1·5 - 1) \exp(\sqrt{1·25}) \approx 1·5294173$

Strip 2   [1·5, 2] $A_2 \approx (2 - 1·5) \exp(\sqrt{1·75}) \approx 1·8771008$

Strip 3   [2, 2·5] $A_3 \approx (2·5 - 2) \exp(\sqrt{2·25}) \approx 2·2408445$

Strip 4   [2·5, 3] $A_4 \approx (3 - 2·5) \exp(\sqrt{2·75}) \approx 2·6252213$

$$\text{TOTAL} \qquad 8·2725839$$

so $A = 8·27$ to 2 d.p.

COMPOUND TRAPEZIUM RULE

This is of great theoretical and practical importance and so we will formulate it in a general manner before we consider any particular example.

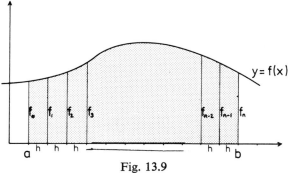

Fig. 13.9

It is helpful to consider the case of n strips defined by the $n+1$ ordinates whose values are $f_0, f_1, f_2, \ldots f_n$ (fig. 13.9) with strip width h in each case.

Applying the Trapezium Rule separately to each strip we get

$$A \approx (h/2) \times (f_0 + f_1) + (h/2) \times (f_1 + f_2) + \ldots + (h/2) \times (f_{n-1} + f_n)$$

$$\therefore \quad A \approx (h/2) \times (f_0 + 2f_1 + 2f_2 + \ldots + 2f_{n-1} + f_n)$$

$$\ldots \underline{\text{COMPOUND TRAPEZIUM RULE}}$$

For our example $n=4$, $a=1$, $b=3$, $\therefore$ $h=(b-a)/n=0\cdot5$

$$A \approx (0\cdot5/2) \times (\exp(\sqrt{1}) + 2\exp(\sqrt{1\cdot5})$$
$$+ 2\exp(\sqrt{2}) + 2\exp(\sqrt{2\cdot5}) + \exp(\sqrt{3}))$$

$\therefore$ $A = 8\cdot28$ to 2 d.p.

Comparing the results ($8\cdot27$ and $8\cdot28$) using the two different compound rules we are aware of much closer consistency now, suggesting that the answer is $8\cdot3$ correct to 2 s.f. However it is preferable to seek consistency by employing *one* rule, often the Compound Trapezium Rule, and applying it with $n=2$, $n=4$, $n=8$, $n=16$ ... strips until the sequence of resulting estimates is seen to converge to some value. Only in unusual circumstances is this likely to be misleading. However, round off errors may ultimately lead to a fall-off in accuracy as n gets very large. The *doubling* of strip size each time is not essential but allows all the previously computed function values to be used again, an important economy of effort. Using equal strip widths is also merely a matter of convenience.

For $\int_1^3 \exp(\sqrt{x})\,dx$ the repeated application of the Compound Trapezium Rule gives the values in Table 13.2.

TABLE 13.2

n	Estimates
1	$8\cdot3705$
2	$8\cdot2985$
4	$8\cdot2811$
8	$8\cdot2769$
16	$8\cdot2758$

This entails a fairly long series of calculations admittedly, but it is straightforward to do.

On the pocket calculator, the procedure can be effected fairly automatically. This will be illustrated by starting with the case $n=4$ and then extending this to $n=8$.

Initial formula

$$A \approx (h/2) \times (f_0 + 2f_1 + 2f_2 + 2f_3 + f_4),$$

$h=0\cdot5$, $f_0 = \exp(\sqrt{1})$, $f_1 = \exp(\sqrt{1\cdot5})$, $f_2 = \exp(\sqrt{2})$, $f_3 = \exp(\sqrt{2\cdot5})$, $f_4 = \exp(\sqrt{3})$.

	Algebraic key strokes	Display	Comment	Reverse Polish key strokes	
(a)	1·5	1·5	x_1	1·5	(a)
(b)	$\boxed{\sqrt{x}}$ $\boxed{e^x}$	3·4032976	f_1	$\boxed{\sqrt{x}}$ $\boxed{e^x}$	(b)
(c)	$\boxed{+}$ 2	2·	x_2	2	(c)
(d)	$\boxed{\sqrt{x}}$ $\boxed{e^x}$	4·1132503	f_2	$\boxed{\sqrt{x}}$ $\boxed{e^x}$	(d)
(e)	$\boxed{+}$ 2·5	2·5	x_3	$\boxed{+}$ 2·5	(e)
(f)	$\boxed{\sqrt{x}}$ $\boxed{e^x}$	4·8604879	f_3	$\boxed{\sqrt{x}}$ $\boxed{e^x}$	(f)
(g)	$\boxed{\times}$ 2 $\boxed{+}$	24·754072	$2(f_1+f_2+f_3)$	$\boxed{+}$ 2 $\boxed{\times}$	(g)
(h)	1	1·	x_0	1	(h)
(i)	$\boxed{\sqrt{x}}$ $\boxed{e^x}$	2·7182818	f_0	$\boxed{\sqrt{x}}$ $\boxed{e^x}$	(i)
(j)	$\boxed{+}$ 3	3·	x_4	$\boxed{+}$ 3	(j)
(k)	$\boxed{\sqrt{x}}$ $\boxed{e^x}$	5·6522336	f_4	$\boxed{\sqrt{x}}$ $\boxed{e^x}$	(k)
(l)	$\boxed{\times}$ ·5 $\boxed{\div}$ 2			$\boxed{+}$ ·5 $\boxed{\times}$ 2	(l)
	$\boxed{=}$	8·2811468	A estimate	$\boxed{\div}$	

The problem is now not to lose the calculations already made. The part required again is $f_0+2f_1+ \ldots +f_4$. If we therefore multiply 8·2811468 (A) by $2/h$ we shall get this back. Next we wish to calculate the extra intermediate function values, and then use $h=0.25$ to complete the new estimate from

$$A \approx (0\cdot25/2) \times (\underset{\substack{\downarrow \\ \text{new} \\ x=1\cdot25}}{\overset{\substack{\text{old} \\ x=1 \\ \downarrow}}{f_0}}+2f_{\frac{1}{2}}+\underset{\substack{\downarrow \\ \text{new} \\ x=1\cdot75}}{\overset{\substack{\text{old} \\ x=1\cdot5 \\ \downarrow}}{2f_1}}+2f_{1\frac{1}{2}}+\overset{\substack{\text{old} \\ x=2 \\ \downarrow}}{2f_2}+\underset{\substack{\downarrow \\ \text{new} \\ x=2\cdot25}}{2f_{2\frac{1}{2}}}+\overset{\substack{\text{old} \\ x=2\cdot5 \\ \downarrow}}{2f_3}+\underset{\substack{\downarrow \\ \text{new} \\ x=2\cdot75}}{2f_{3\frac{1}{2}}}+\overset{\substack{\text{old} \\ x=3 \\ \downarrow}}{f_4})$$

i.e.

$$A \approx (0\cdot25/2) \times (\underbrace{\{f_0+2f_1+2f_2+2f_3+f_4\}}_{\text{ALREADY COMPUTED}} + \underbrace{\{2f_{\frac{1}{2}}+2f_{1\frac{1}{2}}+2f_{2\frac{1}{2}}+2f_{3\frac{1}{2}}\}}_{\text{STILL TO BE COMPUTED}})$$

	Algebraic key strokes	Display	Comment	Reverse Polish key strokes	
(m)	$\boxed{\times}$ 2 $\boxed{\div}$ ·5		f_0+2f_1+	2 $\boxed{\times}$ ·5 $\boxed{\div}$	(m)
	$\boxed{=}$ $\boxed{\text{STO}}$	33·124587	$\ldots +f_4$		
(n)	1·25 $\boxed{\sqrt{x}}$ $\boxed{e^x}$	3·0588345	$f_{\frac{1}{2}}$	1·25 $\boxed{\sqrt{x}}$ $\boxed{e^x}$	(n)
(o)	$\boxed{+}$ 1·75 $\boxed{\sqrt{x}}$			1·75 $\boxed{\sqrt{x}}$ $\boxed{e^x}$	(o)
	$\boxed{e^x}$	3·7542016	$f_{1\frac{1}{2}}$		
(p)	$\boxed{+}$ 2·25 $\boxed{\sqrt{x}}$			$\boxed{+}$ 2·25 $\boxed{\sqrt{x}}$	(p)
	$\boxed{e^x}$	4·481689	$f_{2\frac{1}{2}}$	$\boxed{e^x}$	

(continued)

	Algebraic key strokes	Display	Comment	Reverse Polish key strokes	
(q)	[+] 2·75 [√x] [e^x]	5·2504426	$f_{3\frac{1}{2}}$	[+] 2·75 [√x] [e^x]	(q)
(r)	[×] 2 [+] [RCL] [×]	66·214923	$f_0 + 2f_{\frac{1}{2}} + \ldots + f_4$	[+] 2 [×] [+]	(r)
(s)	·25 [÷] 2 [÷]	8·2768654	A estimate	·25 [×] 2 [÷]	(s)

Notes on Algebraic key-stroke procedure:

(i) If there are parentheses the procedure can be modified to do the multiplication by 2 as soon as the values are computed, and to add them to the 33·124587, thus:

(m)　[×] 2 [÷] ·5 [+] [(]

(n)　1·25 [√x] [e^x] [×] 2 [)]

etc.

(ii) If a 'sum of products' logic machine is used then the calculations $2 \times f_{\frac{1}{2}}$ etc. can be directly performed and added to the 33·124587 thus:

(m)　[×] 2 [÷] ·5

(n)　[+] 1·25 [√x] [e^x] [×] 2

etc.

and line (r) must be modified for such machines:

(r)　[=] [×] 2 [+] [RCL] [=] [×]

(iii) Another procedure is to divide the $f_0 + 2f_1 + \ldots + f_4$ to give $\frac{1}{2}f_0 + f_1 + f_2 + \ldots \frac{1}{2}f_4$, and then add in the extra terms $f_{\frac{1}{2}}, f_{1\frac{1}{2}} \ldots$ and finally multiply by h ($= ·25$) instead of by $h/2$:

(m)　[÷] ·5 [+]　　16·562294　$\frac{1}{2}f_0 + \ldots + \frac{1}{2}f_4$

(n)　1·25 [√x] [e^x]　3·0588346　$f_{\frac{1}{2}}$

etc.

You, the reader, may well be able to simplify the suggested algorithms to suit your style of calculating but the important thing is not to lose track of what you are doing!

Some things to do

Exercise 13.13　Estimate the integral

$$\int_1^2 \frac{1}{1 + \sqrt[5]{x}} \, dx$$

using (a) the Midpoint Rule, (b) The Trapezium Rule.

Exercise 13.14 Repeat Exercise 13.13, using the Compound Trapezium Rule with $n = 2, 4, \ldots$ strips sufficient to give 4 d.p. accuracy as judged by convergence of the sequence of estimates.

Exercise 13.15 The integral

$$\int_0^{x_0} \frac{x^4 e^x}{(e^x - 1)^2} \, dx$$

occurs when finding the heat capacity of a solid using a method based on the vibrational frequencies of the crystal ($x = hf/kT$). Find the value when $x_0 = 1$ using the Trapezium Rule with 4 and 8 strips (assume $\dfrac{x^4 e^x}{(e^x - 1)^2} = 0$ when $x = 0$)

SIMPSON'S RULE

This is another very popular formula for approximating an integral, based on three points rather than on the two for the Trapezium Rule. This time, instead of using a straight line approximation, a parabola is fitted to pass through the three points with ordinate values f_0, f_1, f_2 for a pair of strips (fig. 13.10).

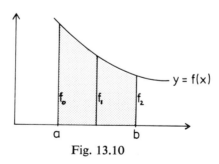

Fig. 13.10

The formula is

$$A \approx (h/3) \times (f_0 + 4f_1 + f_2) \ldots \underline{\text{SIMPSON'S RULE}}$$

Applying it to the integral estimated earlier:

$$\int_1^3 \exp(\sqrt{x}) \, dx \approx \tfrac{1}{3}(\exp(\sqrt{1}) + 4\exp(\sqrt{2}) + \exp(\sqrt{3})) \approx 8 \cdot 2745$$

If we compare this with the results of using the Trapezium Rule with 2 strips we see that we have here a much more accurate result for the same amount of work, i.e. for the same number of function evaluations, since this result is correct to 2 d.p.

COMPOUND SIMPSON'S RULE

We can take n strips (n even) and use Simpson's Rule separately on each pair of strips. The appropriate formula is

$$A \approx (h/3) \times (f_0 + 4f_1 + 2f_2 + 4f_3 + 2f_4 + \ldots + 4f_{n-1} + f_n)$$
$$\ldots \text{COMPOUND SIMPSON'S RULE}$$

A procedure implementing this rule on a calculator can be devised in much the same way as for the Compound Trapezium Rule but if previous results are to be saved and re-used the calculation can become rather involved. For most integral problems it is likely to be quicker to re-evaluate each time or (for complicated problems) to write down the values and re-enter them as needed.

Consider $\displaystyle\int_0^1 \sin \sqrt{x} \, dx$

If we take 4 strips (fig. 13.11) Simpson's Rule is:

$$A \approx (h/3) \times (f_0 + 4f_1 + 2f_2 + 4f_3 + f_4)$$
$$\approx (0\cdot25/3) \times (\sin \sqrt{0} + 4 \sin \sqrt{0\cdot25} + 2 \sin \sqrt{0\cdot5} + 4 \sin \sqrt{0\cdot75} + \sin \sqrt{1})$$

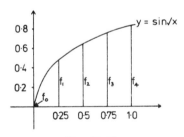

Fig. 13.11

If parentheses or 'sum of products' logic are available then the terms in brackets in this formula can be computed directly on the calculator—as they can of course if using a Reverse Polish logic machine. If none of these facilities is available then a memory can be usefully employed, or some reorganization of the computation is necessary.

(i) *Algebraic key strokes*
     (using memory)   *Display*   *Comment*
     Select radian mode

(a) ·25 $\boxed{\sqrt{x}}$ $\boxed{\sin}$ ·47942554 f_1
(b) $\boxed{+}$ ·75 $\boxed{\sqrt{x}}$ $\boxed{\sin}$ ·76175998 f_3
(c) $\boxed{\times}$ 4 $\boxed{=}$ 4·9647421 $4f_1 + 4f_3$

At this stage it is useful to store the result of $4f_1 + 4f_3$ before calculating the $2f_2$ term.

(d) $\boxed{\text{STO}}$ ·5 $\boxed{\sqrt{\ }}$ $\boxed{\sin}$　　　　·64963694 f_2

(e) $\boxed{\times}$ 2 $\boxed{+}$　　　　　　　1·2992739 $2f_2$

(f) $\boxed{\text{RCL}}$ $\boxed{+}$　　　　　　6·264016 $4f_1 + 2f_2 + 4f_3$

(g) 0 $\boxed{\sqrt{x}}$ $\boxed{\sin}$　　　　　0·　　　　f_0

(h) $\boxed{+}$ 1 $\boxed{\sqrt{x}}$ $\boxed{\sin}$ $\boxed{\times}$　7·1054869 $f_0 + 4f_1 + 2f_2 + 4f_3 + f_4$

(i) ·25 $\boxed{\div}$ 3 $\boxed{=}$　　　　·59212391 estimate of integral

Actually the use of the memory is quite unnecessary if a little ingenuity is employed.

If we calculate $2f_1 + 2f_3$ (rather than $4f_1 + 4f_3$) then we can add f_2 to these giving $2f_1 + f_2 + 2f_3$ and then multiply *all* this by 2 again giving $4f_1 + 2f_2 + 4f_3$ as required, to which f_0 and f_4 are added as before. If this ploy is adopted we get the following key-stroke sequence instead.

(ii) *Algebraic key strokes*
　　　(*not using memory*)　　　　　*Display*　　　　*Comment*
　　　Select radian mode

(a) ·25 $\boxed{\sqrt{x}}$ $\boxed{\sin}$　　　　　·47942554 f_1

(b) $\boxed{+}$ ·75 $\boxed{\sqrt{x}}$ $\boxed{\sin}$　　·76175998 f_3

(c) $\boxed{\times}$ 2 $\boxed{+}$　　　　　　　2·482371 $2f_1 + 2f_3$

(d) ·5 $\boxed{\sqrt{x}}$ $\boxed{\sin}$ $\boxed{\times}$　　3·132008 $2f_1 + f_2 + 2f_3$

(e) 2 $\boxed{+}$　　　　　　　　　6·264016 $4f_1 + 2f_2 + 4f_3$

(f) 0 $\boxed{\sqrt{x}}$ $\boxed{\sin}$　　　　　0·　　　　f_0

(g) $\boxed{+}$ 1 $\boxed{\sqrt{x}}$ $\boxed{\sin}$ $\boxed{\times}$　7·1054869 $f_0 + 4f_1 + 2f_2 + 4f_3 + f_4$

(h) ·25 $\boxed{\div}$ 3 $\boxed{=}$　　　　·59212391 estimate of integral

(iii) *Reverse Polish key strokes*　　*Display*　　　　*Comment*

(a) 0 $\boxed{\sqrt{x}}$ $\boxed{\sin}$　　　　　0·　　　　f_0, included for completeness here

(b) ·25 $\boxed{\sqrt{x}}$ $\boxed{\sin}$　　　　·47942554 f_1

(c) 4 $\boxed{\times}$ $\boxed{+}$　　　　　　1·9177022 $f_0 + 4f_1$

(d) ·5 $\boxed{\sqrt{x}}$ $\boxed{\sin}$　　　　·64963694 f_2

(e) 2 $\boxed{\times}$ $\boxed{+}$　　　　　　3·216976 $f_0 + 4f_1 + 2f_2$

(f) ·75 $\boxed{\sqrt{x}}$ $\boxed{\sin}$　　　　·76175998 f_3

(g) 4 $\boxed{\times}$ $\boxed{+}$　　　　　　6·264016 $f_0 + 4f_1 + 2f_2 + 4f_3$

(h) 1 $\boxed{\sqrt{x}}$ $\boxed{\sin}$　　　　　·84147098 f_4

(i) $\boxed{+}$　　　　　　　　　　7·1054869 $f_0 + 4f_1 + 2f_2 + 4f_3 + f_4$

(j) ·25 $\boxed{\times}$ 3 $\boxed{\div}$　　　　·59212391 estimate of integral

General procedure for n strips using Compound Simpson's Rule:

(i) Compute: $f_1 + f_3 + f_5 + \ldots + f_{n-1}$

(ii) multiply by 2: $2f_1 + 2f_3 + \ldots + 2f_{n-1}$

(iii) add in the values $f_2 + f_4 + f_6 + \ldots + f_{n-2}$ to give
$2f_1 + f_2 + 2f_3 + \ldots + f_{n-2} + 2f_{n-1}$

(iv) multiply by 2 and add f_0 and f_n to give
$f_0 + 4f_1 + 2f_2 + 4f_3 + \ldots 2f_{n-2} + 4f_{n-1} + f_n$

(v) multiply by h and divide by 3 to give the answer
$(h/3) \times (f_0 + 4f_1 + \ldots + f_n)$

Simpson's Rule is generally more accurate than the Trapezium Rule, as is illustrated for the evaluation of $\int_0^3 e^{x^2} \, dx$ (Table 13.3), which requires many strips as $y = e^{x^2}$ has a rapidly changing gradient (check this by calculating a few values of $y = e^{x^2}$). The correct value is 1444·5451 to 8 s.f.

TABLE 13.3

n	Simpson	n	Trapezium
8	1565·9919	32	1479·9375
16	1456·1594	64	1453·4338
32	1445·3784	128	1446·7698
64	1444·5992	256	1445·1015
128	1444·5485	512	1444·6842

Programmable calculators will perform such calculations automatically, of course, and in fact it is possible to buy non-programmable machines (e.g. Commodore SR4190R) which have a key for numerical quadrature, using the Compound Trapezium Rule.

Something to do

Exercise 13.16   Evaluate

$$\int_0^{10} \frac{\sin x}{x} \, dx$$

using the Compound Simpson's Rule with 2, 4, 8 . . . strips to give three decimal place accuracy in the answer. Note that $(\sin x)/x$ approaches the value 1 as x tends to zero, so $f_0 = 1$.

This integral

$$Si(x) = \int_0^x \frac{\sin x}{x} \, dx$$

is known as the Sine Integral, and occurs in certain branches of Physics (e.g. diffraction theory).

RELATIONSHIP BETWEEN STRIP WIDTH AND ERROR

It is obvious from Table 13.3 that every time the strip width is halved (and of course the number of strips n doubled) the error in the estimate is reduced.

Given that the correct answer is 1444·5451 to 8 s.f. complete Table 13.4 below which indicates the errors for different numbers of strips.

TABLE 13.4

No. of strips	Errors in Trapezium estimates
n	$E_t(n)$
32	35·3924
64	. . . . . .
128	. . . . . .
256	0·5564
512	. . . . . .

Is there any pattern to the terms $E_t(n)$?

Calculate the ratios, $\dfrac{E_t(64)}{E_t(32)}$ and $\dfrac{E_t(128)}{E_t(64)}$ and $\dfrac{E_t(256)}{E_t(128)}$ and $\dfrac{E_t(512)}{E_t(256)}$.

What do you notice?

You should see that each halving of h (doubling of n) leads to the error in the estimate being reduced to about $\frac{1}{4}$ its former level. This suggests a square law relationship, because $(\frac{1}{2})^2 = \frac{1}{4}$. This can be expressed by saying that E_t is proportional to h^2, so that if h is changed to $h/2$, E_t is then proportional to $h^2/4$, i.e. $\frac{1}{4}$ of what it was before.

(*Note*: the relationship is not strictly this but becomes more and more accurate as h is reduced).

Some things to do

Exercise 13.17   Draw up a table, similar to Table 13.4, for the Compound Simpson's Rule errors, E_s, in evaluating $\int_0^3 e^{x^2}\, dx$ from the figures provided in Table 13.3. By computing the ratios of successive terms $E_s(2n)/E_s(n)$ deduce that, for small enough h, E_s is proportional to h^4 by noting that the ratios get successively closer to $\frac{1}{16}$ ($=(\frac{1}{2})^4$).

Exercise 13.18   Use the compound Trapezium Rule to evaluate

$$\int_1^2 \frac{1}{1+\sqrt[4]{x}}\, dx$$

taking 2, 4, 8 . . . strips to give a result accurate to 4 d.p. (remember that $\boxed{\sqrt{x}}$ $\boxed{\sqrt{x}}$ will give the necessary fourth root).

Write down the ratio of successive errors and check that the errors are reduced by a factor of about $\frac{1}{4}$ each time.

Exercise 13.19   Use the Compound Simpson's Rule to evaluate

$$\int_1^2 e^x \ln (x)\, dx$$ taking 2, 4, 8, . . . strips. Show that the errors reduce

by about a factor of $\frac{1}{16}$ each time.

HIGHER ORDER FORMULAE

The Trapezium Rule and Simpson's Rule are examples of what are called Newton-Cotes quadrature formulae. As has been mentioned the Trapezium Rule uses a straight line (linear curve) to approximate to the true curve and Simpson's Rule uses a parabola (quadratic curve). Higher order Newton-Cotes formulae can be derived which use cubic, quartic, quintic, . . . curves passing through 4, 5, 6 . . . points respectively. These are not in practice used a great deal as more accurate and more powerful methods for use on computers have been derived in recent times.

The commoner Newton-Cotes formulae are provided in Table 13.5.

TABLE 13.5

Order	Name	Formula
1	Trapezium	$(h/2) \times (f_0 + f_1)$
2	Simpson	$(h/3) \times (f_0 + 4f_1 + f_2)$
3	Three-Eighths	$(3h/8) \times (f_0 + 3f_1 + 3f_2 + f_3)$
4	Boole	$(2h/45) \times (7f_0 + 32f_1 + 12f_2 + 32f_3 + 7f_4)$
5	Six-Point Rule	$(5h/288) \times (19f_0 + 75f_1 + 50f_2 + 50f_3 + 75f_4 + 19f_5)$
6	Seven-Point Rule	$(h/140) \times (41f_0 + 216f_1 + 27f_2 + 272f_3 + 27f_4 + 216f_5 + 41f_6)$

Some things to do

Exercise 13.20   The integral

$$\int_{x_1}^{x_2} \frac{1}{x(e^{hc/xkT} - 1)}\, dx$$

occurs when finding the fraction of total energy that emanates as visible radiation from a black body.

Use Boole's Rule (4 strips) and the Six-Point Rule (5 strips) to evaluate this integral given that

$$hc/kT = 1.41, \quad x_1 = 1.7, \quad x_2 = 2.7$$

Exercise 13.21   The solutions of certain scientific problems, e.g. fluid flow, gravitational potential, lead to what are called *elliptic integrals*. The exact perimeter of an ellipse $x^2/a^2 + y^2/b^2 = 1$ (or in polar co-ordinates $L/r = 1 + e \cos \theta$) is given by the elliptic integral:

$$4a \int_0^{\pi/2} \sqrt{(1 - e^2 \cos^2 \theta)} \, d\theta \quad \text{where } e^2 = 1 - b^2/a^2$$

An *approximate* formula for the perimeter was stated and used on page 46.

Investigate the accuracy of that approximate formula by comparing its results with those from evaluating the above integral by quadrature methods. For various values of eccentricity e draw up tables of values for $f(x) = \sqrt{(1 - e^2 \cos^2 \theta)}$ as below:

x	0	$\pi/12$	$\pi/6$	$\pi/4$	$\pi/3$	$5\pi/12$	$\pi/2$
$f(x)$							

Evaluate the integral using the values from the table,
(*a*) by using the Compound Simpson's Rule,
(*b*) by using the Three-Eighths Rule twice, for $[0, \pi/4]$ and then for $[\pi/4, \pi/2]$,
(*c*) by using the Seven-Point Rule.

Exercise 13.22   A botanist wishes to estimate the volume above ground of a tree trunk up to the first branches. He measures the diameter, d, at 0.5 m intervals and prepares a table of values. Noting that the tree trunk is fairly circular he can easily compute the cross-sectional areas using $\pi d^2/4$. To find the requisite volume he resorts to the Compound Trapezium Rule. What result does he get?

Height, m	0	0·5	1	1·5	2	2·5	3
Average Diameter, m	1·12	0·94	0·83	0·71	0·68	0·64	0·57
Cross-sectional Area, m²							

Exercise 13.23   In gas chromatographic analysis the areas under the peaks of the curve produced by the pen-recorder or digital recorder are proportional to the amounts of various chemicals present in the gas being analysed. Use an appropriate numerical method to compare the amounts of the two chemicals indicated by the following table of values measured from a chromatogram:

Retention time	x	2·2	2·4	2·6	2·8	3·0	3·2	3·4	3·6
Detector response	y	0	0·12	0·63	1·96	1·83	0·74	0·41	0·10

	x	3·8	...	...	...	5·2	5·4	5·6	5·8
	y	0	...	...	...	0	0·21	0·73	1·29

	x	6·0	6·2	6·4	6·6	6·8	7·0	7·2	7·4
	y	1·54	1·09	0·79	0·54	0·27	0·11	0·06	0

Exercise 13.24   There is a fairly straightforward quadrature technique, called Romberg Integration, which can give much more accurate results, using the Compound Trapezium Rule, for example. Find out about it and repeat some of the problems in this chapter.

14. Case studies in physics

CASE STUDY 1:

Two dimensional motion—Satellite orbits

In the case of a satellite orbiting the Earth the force acting on the satellite is directed towards the centre of the Earth and with a suitable choice of units, is given by the equation $F = -m/r^2$ (so $a = -1/r^2$) where m is the satellite's mass and r is its distance from the Earth's centre (fig. 14.1).

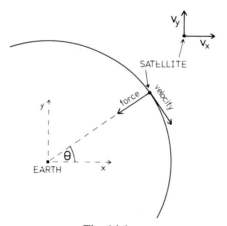

Fig. 14.1

At time t the satellite is at (x, y) with velocity components (v_x, v_y) and acceleration components $(-1/r^2 \cos \theta, -1/r^2 \sin \theta)$. To find the new position (X, Y) and velocity (V_x, V_y) at time Δt later we use:

$$V_x = v_x - \left(\frac{1}{r^2} \cos \theta\right) \Delta t, \quad V_y = v_y - \left(\frac{1}{r^2} \sin \theta\right) \Delta t$$

Now

$$r^2 = x^2 + y^2, \quad \cos \theta = \frac{x}{\sqrt{(x^2 + y^2)}}, \quad \sin \theta = \frac{y}{\sqrt{(x^2 + y^2)}}$$

so we get

$$V_x = v_x - \frac{x\Delta t}{(x^2 + y^2)^{3/2}} \qquad V_y = v_y - \frac{y\Delta t}{(x^2 + y^2)^{3/2}}$$

and we have everything we need to find the new velocity components—we already know x, y, v_x, v_y, and we choose Δt as some small value.

To find the new position, which really means finding the *difference* in positions at t and at $t + \Delta t$, we simply need

$$X = x + v_x\Delta t, \quad Y = y + v_y\Delta t$$

Example

We take the position of the Earth as $(0, 0)$ and the initial position of the satellite as $(0, 1)$ with velocity $(0 \cdot 8, 0)$. We will take $\Delta t = 0 \cdot 5$, so we have:

$$v_x = 0 \cdot 8, \quad v_y = 0, \quad x = 0, \quad y = 1, \quad \Delta t = 0 \cdot 5$$

$$V_x = 0 \cdot 8 - \frac{0 \times 0 \cdot 5}{(0^2 + 1^2)^{3/2}} = 0 \cdot 8, \quad V_y = 0 - \frac{1 \times 0 \cdot 5}{(0^2 + 1^2)^{3/2}} = -0 \cdot 5$$

$$X = 0 + 0 \cdot 8 \times 0 \cdot 5 = 0 \cdot 4, \qquad Y = 1 + 0 \times 0 \cdot 5 = 1$$

We have now found the position and velocity of the satellite at $0 \cdot 5$ time units later—it is at $(0 \cdot 4, 1)$ travelling with velocity $(0 \cdot 8, -0 \cdot 5)$.

Obviously a calculator is invaluable for accurately and speedily performing the arithmetic. We could now repeat the computations for another time step starting at $t = 0 \cdot 5$, with the new starting conditions $v_x = 0 \cdot 8$, $v_y = -0 \cdot 5$, $x = 0 \cdot 4$, $y = 1$, $\Delta t = 0 \cdot 5$.

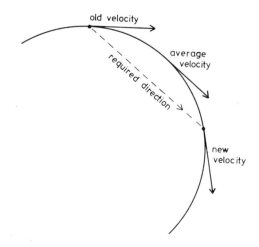

Fig. 14.2

However, the above very simple procedure has an inherent weakness. We are always using the old velocity components (v_x, v_y) to calculate the new X and Y values. To use the new values (V_x, V_y) would not be much better but if we use the *average* of the two $((v_x + V_x)/2, (v_y + V_y)/2)$ then we get a much more representative value, far more likely to keep us on the correct orbital path (see fig. 14.2).

Thus we calculate $V_x = 0.8$ and $V_y = -0.5$ just as before, but then calculate

$$V_{xav} = \frac{v_x + V_x}{2} = \frac{0.8 + 0.8}{2} = 0.8, \quad V_{yav} = \frac{v_y + V_y}{2} = \frac{0 + -0.5}{2} = -0.25$$

Now we can find a rather better estimate for the position of the satellite after time 0.5:

$$X = 0 + 0.8 \times 0.5 = 0.4$$

$$Y = 1 - 0.25 \times 0.5 = 0.875$$

So the satellite at time $t = 0.5$ is at position $(0.4, 0.875)$.

In future calculations we use the average velocity of the previous step $[(0.8, -0.25)$ in this case$]$ and when we use the equations for V_x and V_y we will automatically get the average velocity for the step we are currently interested in.

We now set out the key strokes for the first step, and then for the general step:

$$x = 0, \quad y = 1, \quad v_x = 0.8, \quad v_y = 0, \quad \Delta t = 0.5$$

(continued)

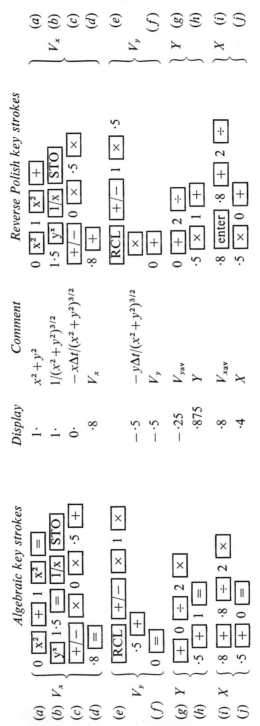

	Algebraic key strokes	Display	Comment	Reverse Polish key strokes
(a) V_x	0 x^2 + 1 x^2 = ×	1·	$x^2 + y^2$	0 x^2 1 x^2 +
(b)	y^x 1·5 = 1/x STO	1·	$1/(x^2+y^2)^{3/2}$	1·5 y^x 1/x STO
(c)	+/− × 0 × ·5 +	0·	$-x\Delta t/(x^2+y^2)^{3/2}$	+/− 0 × ·5 ×
(d)	·8 =	·8	V_x	·8 +
(e) V_y	RCL +/− × 1 ×	−·5	$-y\Delta t/(x^2+y^2)^{3/2}$	RCL +/− 1 × ·5
(f)	·5 +	−·5	V_y	0 +
	0 =			
(g) Y	+ 0 ÷ 2 ×	−·25	V_{yav}	0 + 2 ÷
(h)	·5 × 1 =	·875	Y	·5 × 1 +
	+			
(i) X	·8 + ·8 ÷ 2 ×	·8	V_{xav}	·8 enter ·8 + 2 ÷
(j)	·5 + 0 =	·4	X	·5 × 0 +

Now for all subsequent steps the procedure is very similar. Lines (a) to (f) are exactly the same, using the new data of course, and (g) to (j) are reduced as the averaging need no longer be done.

	Algebraic key strokes	Display	Comment	Reverse Polish key strokes	
(a)	·4 [x²] [+] ·875 [x²] [=]	·925625	$x^2 + y^2$	·4 [x²] ·875 [x²] [+]	(a)
(b)	[yˣ] 1·5 [=] [1/x] [STO]	1·1229163	$1/(x^2+y^2)^{3/2}$	1·5 [yˣ] [1/x] [STO]	(b)
(c)	[+/−] [×] ·4 [×] ·5 [+]	−·22458326	$-x\Delta t/(x^2+y^2)^{3/2}$	[+/−] ·4 [×] ·5 [×]	(c)
(d)	·8 [=]	·57541674	V_x	·8 [+]	(d)
(e)	[RCL] [+/−] [×] ·875	−·49127588	$-y\Delta t/(x^2+y^2)^{3/2}$	[RCL] [+/−] ·875 [×]	(e)
(f)	[×] ·5 [−] ·25 [=]	−·74127588	V_y	·5 [×] ·25 [−]	(f)
(g)	[×] ·5 [+] ·875 [=]	·50436206	$Y = V_y\Delta t + y$	·5 [×] ·875 [+]	(g)
(h)	·5754 [×] ·5 [+] ·4 [=]	·6877	$X = V_x\Delta t + x$	·5754 [enter] ·5 [×] ·4 [+]	(h)

Lines (a)–(d) give V_x; lines (e)–(f) give V_y; line (g) gives Y; line (h) gives X.

We now have $x = 0.4$, $y = 0.875$, $V_x = 0.8$, $V_y = -0.25$, $t = 0.5$.

We have now found the position of the satellite at time $t = 1$ to be $(0.6877, 0.5044)$, and we could continue step by step until the orbital path is clear.

Continuing we would get the results shown in Table 14.1, figures finally rounded to 2 d.p.

TABLE 14.1

Time	X	Y	Average V_x	Average V_y
0	0.00	1.00	0.80	0.00
0.5	0.40	0.88	0.80	-0.25
1	0.69	0.50	0.58	-0.74
1.5	0.70	-0.07	0.00	-1.15
2	0.20	-0.59	-0.99	-1.05
2.5	-0.50	-0.52	-1.40	0.16
3	-0.86	-0.01	-0.73	0.86
3.5	-0.89	0.38	-0.01	0.92
4	-0.67	0.73	0.43	0.72
4.5	-0.29	0.91	0.78	0.34
5	0.19	0.81	0.94	-0.19

By time $t = 5$ we have entered the second revolution. A graph of these results is shown in fig. 14.3.

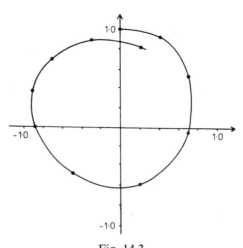

Fig. 14.3

The general shape of the orbit is clear but it is not quite the closed ellipse it should be. Why is this? The errors in using such a crude approximate method as we have adopted with such a *large* time step Δt (0·5) have accumulated. The only way to get greater accuracy with this method is to reduce the Δt to 0·1 or 0·05 say—but then the calculations begin to get rather tedious by calculator, and a programmable calculator or computer (if available!) would be appropriate. For example, using $t=0·05$ the values produced very closely correspond to a closed ellipse (fig. 14.4) and the time for one complete revolution (the 'period') is almost exactly 4 as shown in columns 1 to 3 of Table 14.2 which reproduces results for every tenth point.

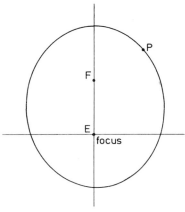

Fig. 14.4

TABLE 14.2

Time	x	y	Distance from E (0, 0)	Distance from F (0, 0·5)	Sum
0	0·00	1·00	1·00	0·50	1·50
0·5	0·38	0·87	0·95	0·53	1·48
1	0·65	0·50	0·82	0·65	1·47
1·5	0·61	−0·08	0·62	0·84	1·46
2	−0·03	−0·47	0·47	0·97	1·44
2·5	−0·63	−0·04	0·63	0·83	1·46
3	−0·65	0·53	0·84	0·65	1·49
3·5	−0·37	0·89	0·93	0·54	1·47
4	0·02	1·00	1·00	0·50	1·50

A good test that a curve is an ellipse is to check that whatever point P on the curve we choose, the sum of the distances from P to the two foci is constant. Now we know that the Earth (0, 0) is one focus. Where is the other? By symmetry it must be on the y-axis and at about $y=0·5$.

It is now a simple matter to compute the necessary sums of distances which turn out to be fairly consistent. The distance of a point P with coordinates (x, y) from another point A, coordinates (a, b) is given by $\sqrt{((x-a)^2+(y-b)^2)}$.

Distance from E $(0, 0)$ is simply $\sqrt{(x^2+y^2)}$. Distance from F $(0, 0\cdot5)$ is $\sqrt{(x^2+(y-0\cdot5)^2)}$.

$$\therefore \quad \text{Sum} = \sqrt{(x^2+y^2)}+\sqrt{(x^2+(y-0\cdot5)^2)}.$$

Having established Kepler's First Law, namely that the orbit is an ellipse with the Earth's centre at one focus, we can seek to show the Second Law, that the imaginary radial line from the Earth to the satellite sweeps out equal areas in equal times (fig. 14.5).

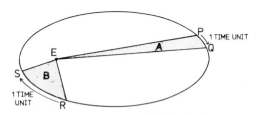

Fig. 14.5   Kepler's Second Law: Area $A = $ Area B.

We have, in Table 14.2, the positions at equal time intervals, so we just need to find the area swept out for each $0\cdot5$ time unit.

In order to simplify finding such an area we might assume the shape to be triangular. This will always underestimate the area, more so for positions nearer the Earth, but for reasonably small time intervals will be satisfactorily consistent.

The formula for the area of the triangle joining three points $E(x_3, y_3)$ and $P(x_1, y_1)$ and $Q(x_2, y_2)$ is

$$\tfrac{1}{2}[x_1(y_2-y_3)+x_2(y_3-y_1)+x_3(y_1-y_2)]$$

In our case $x_3=0$, $y_3=0$, so Area $=\tfrac{1}{2}[x_1y_2-x_2y_1]$. This is easily evaluated on any calculator—the best procedure will depend on* the particular machine.

For $P=(0, 1)$ and $Q=(0\cdot38, 0\cdot87)$:

Area $A \approx \tfrac{1}{2}[0 \times 0\cdot87 - 1 \times 0\cdot38] = -0\cdot19$ (the negative sign can be ignored)
For $R=(-0\cdot03, -0\cdot47)$ and $S=(-0\cdot63, -0\cdot04)$

area $B \approx \tfrac{1}{2}[(-0\cdot03 \times -0\cdot04)-(-0\cdot63 \times -0\cdot47)] = -0\cdot15$.

The discrepancy between these results is mainly due to the relatively large time interval taken or the inappropriateness of the simple triangle area formula. A more accurate formula is presented in Exercise 14.2 below. (Some different approaches to satellite problems using a pocket calculator are provided by E. Mendoza (1975), D. S. Chandler (1976) and R. F. Phillips (1977).)

Some things to do

Exercise 14.1
- (a) Find the paths of the two satellites orbiting the Earth [at (0, 0)] which have velocities of
  - (i) 0·9 unit, (ii) 1 unit, in the *x*-direction, when passing through the point (0, 1).
- (b) What is special about the velocities and the orbit in *a* (ii)?
- (c) Estimate the periods and semi-major axes of the satellites' orbits.
- (d) Check whether Kepler's First and Second Laws are obeyed— i.e. are the paths elliptic and are equal areas swept out in equal times?

Exercise 14.2   A more accurate formula for estimating the area swept out by a satellite when travelling from $P(x_1, y_1)$ to $Q(x_2, y_2)$ is

$$\text{Area} \approx \tfrac{1}{2}(x_1 y_2 - x_2 y_1) \times \left[1 + \frac{\{(x_1 - x_2)^2 + (y_1 - y_2)^2\}}{\{\sqrt{(x_1^2 + y_1^2)} + \sqrt{(x_2^2 + y_2^2)}\}^2} \right]$$

For the data of Table 14.2 compute the area estimates
(i) with the simpler formula $\tfrac{1}{2}[x_1 y_2 - x_2 y_1]$,
(ii) using the improved formula.

Exercise 14.3   Kepler's Third Law states that the period squared is proportional to the semi-major axis cubed. I.e. $\tau^2 \propto a^3$. For the '0·8 velocity' orbit we saw that the period, τ, is 4. The semi-major axis is simply half the longest diameter of the ellipse—in this case the diameter lies on the *y* axis and is 1·5 units, so $a = 0.75$.

Thus we should have $\dfrac{\tau^2}{a^3} = \text{constant}: \dfrac{4^2}{0.75^3} \approx 37.9.$

The following table has been drawn up from the results of accurately computing the orbits of satellites with various velocities. Test the validity of Kepler's Third Law (i.e. see if τ^2/a^3 is the same constant for each case). What is the most economical way to perform the necessary calculations?

Orbital Velocity at (1, 0)	Period, τ	Semi-major axis, a
0·6	2·99	0·61
0·7	3·38	0·66
0·8	3·96	0·735
0·9	4·84	0·84
1·0	6·28	1·00
1·1	8·94	1·26
1·2	14·99	1·78
1·3	36·40	3·23

CASE STUDY 2:

Motion through a resisting medium

For a spherical object of radius r (metres) falling through the air it is known that the resistive force R (newtons) is given by:

$$R = 3 \cdot 1 \times 10^{-4} rv + 0 \cdot 87 r^2 v^2$$

(If the term in v^2 can be neglected this reduces to the familiar Stokes' equation.)

The equation of motion (writing dv/dt for a) is

$$m(dv/dt) = mg - 3 \cdot 1 \times 10^{-4} rv - 0 \cdot 87 r^2 v^2 \ldots \text{I}$$

We will now solve this differential equation numerically to find the velocity—time graph for a water drop of radius 3 mm and density $\rho = 1000$ kg m^{-3}. (In practice the radius r would vary as the drop falls but we shall ignore this.)

$$m = \tfrac{4}{3}\pi r^3 \rho = \tfrac{4}{3}\pi \times (3 \times 10^{-3})^3 \times 1000 \approx 1 \cdot 1310 \times 10^{-4} \text{ kg}$$

From the differential equation I we can get the following expression for the acceleration a $(= dv/dt)$:

$$a \approx 9 \cdot 81 - \frac{3 \cdot 1 \times 10^{-4} \times 3 \times 10^{-3} v}{1 \cdot 131 \times 10^{-4}} - \frac{0 \cdot 87 \times (3 \times 10^{-3})^2 v^2}{1 \cdot 131 \times 10^{-4}}$$

$$a \approx 9 \cdot 81 - 0 \cdot 00822 v - 0 \cdot 0692 v^2 \ldots \text{II}$$

We wish to plot the graph of v against t.

If we use $v_{n+1} = v_n + a\Delta t$ we have a means to find a new velocity v_{n+1} given the previous velocity v_n, knowing the acceleration a for some small time interval Δt.

If we make Δt small enough then we can approximately say that the acceleration, a, will be constant over the time interval. Now in formula II we have an expression for a, namely $9 \cdot 81 - 0 \cdot 00822 v - 0 \cdot 0692 v^2$.

Taking time steps of 0·5 s, with $t = 0$, $v_0 = 0$ initially, we have, after 0·5 s:

$$v = 0 + a \times 0 \cdot 5$$

What is a? We will assume it to be the same as at the *beginning* of the time interval, i.e. at $t = 0$. Then $v = 0$ so $a = 9 \cdot 81$ from formula II.

$$\therefore \quad v = 0 + 9 \cdot 81 \times 0 \cdot 5 = 4 \cdot 905.$$

We now have, at time $t = 0 \cdot 5$, a velocity of 4·905 ms^{-1}. Applying $v_2 = v_1 + a\Delta t$ with $v_1 = 4 \cdot 905$ now and $\Delta t = 0 \cdot 5$ again:

$$v_2 = 4 \cdot 905 + a \times 0 \cdot 5$$

This time formula II is used to calculate a with $v = 4.905$ which is the velocity at the start of the new time interval:

$$v_2 = 4.905 + (9.81 - 0.00822 \times 4.905 - 0.0692 \times 4.905^2) \times 0.5$$
$$\therefore \quad v_2 = 8.9573982$$

The calculator sequence used will depend upon the facilities available on the particular machine being used.

The formula used each time is

$$v_{n+1} = v_n + [9.81 - 0.00822v_n - 0.0692v_n^2] \times \Delta t$$

The values obtained (which the reader can check) are:

Time	0·0	0·5	1·0	1·5	2·0	2·5	3·0	3·5	4·0
v	0·00	4·91	8·96	11·05	11·68	11·82	11·84	11·85	11·85

Very quickly it seems—in about $2\frac{1}{2}$ seconds—the raindrop has virtually attained its 'terminal velocity' of about 11.85 ms^{-1} and it continues to fall at this constant speed.

The velocity–time graph can be sketched as in fig. 14.6.

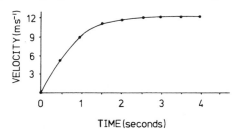

Fig. 14.6   Velocity–time graph of a falling raindrop.

The method we have just outlined is known as Euler's method. Its weakness is that very many small steps are needed for most problems to get accurate results.

Improving Euler's Method

One reason why Euler's method is often rather poor is that it uses the *initial* value as the average for the *whole* step. In our example we were interested in the acceleration given by

$$a = 9.81 - 0.00822v - 0.0692v^2$$

For the first time interval $t = 0$ to $t = 0.5$ we used $v_1 = v_0 + a\Delta t$ taking a to be the value when $t = 0$. It would be better to have taken the average value over the whole time interval, but that is unknown. Perhaps the value of a at the half-way mark, $t=0.25$ would do. This also is unknown. Perhaps the average of a at $t=0$ and a at $t=0.5$ will do as an estimate for it. To get this we need to know v at $0(v_0)$ and v at $0.5(v_1)$ and of course we are trying to *find* v_1! All is not lost, however. We can use the ordinary Euler estimate for v to give an approximate value for a at $t=0.5(a_1)$ and then calculate $a_{av}=(a_0+a_1)/2$ and use a_{av}.

Thus the procedure to find v_{n+1}, knowing v_n and a_n is as follows:

(*a*) Use Euler's method to estimate v_{n+1} $(=v_n+a_n\Delta t)$

(*b*) Use v_{n+1} to estimate a_{n+1} $(=9.81-0.00822v_{n+1}-0.0692v_{n+1}{}^2)$

(*c*) Find the average of a_n and a_{n+1}, a_{av} $\left(=\dfrac{a_n+a_{n+1}}{2}\right)$

(*d*) Use a_{av} to compute v_{n+1} more accurately $(=v_n+a_{av}\Delta t)$.

Example

$\Delta t=0.5$, $v_0=0$ and $a_0=9.81$ are known

(*a*) Euler's method: $v_1=0+9.81\times0.5=4.905$

(*b*) $a_1=9.81-0.00822\times4.905-0.0692\times4.905^2=8.1047964$

(*c*) $a_{av}=\dfrac{9.81+8.1047964}{2}=8.9574$

(*d*) $v_1=0+8.9574\times0.5=4.4787$

v_1 is now known and we carry on to the next time-step. . . .

Greater computational effort is needed for each step compared with the basic Euler method but since much larger step sizes can usually be employed for the same accuracy the method is indeed generally an improvement. Interestingly enough this is not particularly true for the raindrop problem which is something of an exception because there the errors in the Euler method tend to cancel each other out rather than build up. Table 14.3 sets out some results for comparison which the reader might check.

TABLE 14.3

Method	Δt	0	0.5	1	1.5	2	2.5
Euler	0.5	0	4.91	8.96	11.05	11.68	11.82
Improved Euler	0.5	0	4.48	7.71	9.63	10.68	11.24
Analytic	—	0	4.65	8.06	10.05	11.06	11.53

Method	Δt	3	3.5	4	4.5	5
Euler	0.5	11.84	11.85	11.85	11.85	11.85
Improved Euler	0.5	11.53	11.69	11.76	11.80	11.83
Analytic	—	11.74	11.83	11.87	11.89	11.90

Some things to do

Exercise 14.4 A spherical pebble of radius 0·9 cm, and density 2400 kg m^{-3} drops from an aircraft. Plot its velocity–time graph at time intervals of 1 s using the Euler method. Estimate the terminal velocity.

Exercise 14.5

(a) The terminal velocity can be found be setting $dv/dt=0$—this is when there is no resultant acceleration.

For the raindrop and pebble examples this requires the solving of a quadratic equation. Find the terminal velocities and compare with the values from the solutions of the differential equations.

(b) Try to solve the raindrop problem analytically.

Exercise 14.6 The improved Euler Method outlined in the text uses first the basic Euler Method to estimate v_{n+1} (step (a)) and then 'corrects' this first estimate, just once, using a_{av} (steps (b), (c), (d)). It is quite possible to take this corrected v_{n+1} and *re-correct* it (repeating steps (b), (c), (d)) . . . and so on until no further change in v_{n+1} is observed. This method—using several iterations instead of one iteration—gives better results. Investigate its application to the raindrop problem.

Exercise 14.7 Workers in chemical kinetics often establish mathematical models involving differential equations. One such equation, whose solution cannot be expressed in the form $y=F(t)$, arose in enzyme kinetics.

$$\frac{dy}{dt}=\frac{1}{y}-1; \quad y=2 \text{ when } t=0$$

Solve this by the improved Euler method, taking time steps $\Delta t=1$ for $t=0$ to $t=5$.

Exercise 14.8 The difference equations studied in Chapter 12 can often be replaced by differential equations. The logistic model

$$N_{k+1}=N_k+BN_k-dN_kN_k$$

can be formulated as

$$N_{k+1}-N_k=(B-dN_k)N_k; \quad N_1=80$$

and the differential equation equivalent to this is

$$\frac{dN}{dt}=(B-dN)N; \quad N=80 \text{ at } t=0.$$

Solve this differential equation for $B=0·3$, $d=0·0018$ and compare with Table 12.2 (page 204).

An appropriate method is Euler, with $\Delta t=1$, or it can be done analytically.

CASE STUDY 3:

An Electric Circuit

When a charged capacitor of capacitance C farads is connected across a resistor of resistance R ohms (fig. 14.7), then the charge, q coulombs, on the capacitor plate changes according to

$$\frac{dq}{dt} = \frac{-q}{RC}$$

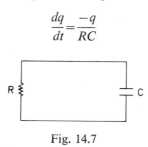

Fig. 14.7

If the initial charge (at $t=0$) is q_0 then this differential equation has a solution $q = q_0 e^{(-t/RC)}$.

This is the familiar exponential decay model which is common in scientific problems.

For the case $R = 10^6\ \Omega$, $C = 2 \times 10^{-6} F$ with $q_0 = 2 \times 10^{-6}$ coulombs the differential equation is

$$\frac{dq}{dt} = -0\cdot5q; \quad q_0 = 2 \times 10^{-6}.$$

Something to do

Exercise 14.9

   (*a*) Solve this differential equation by the ordinary Euler method. What mathematical model does the Euler method reduce the problem to? [Take $\Delta t = 0\cdot5$ and solve for $t=0$ to $t=5$.]

   (*b*) For comparison solve by the Improved Euler method using a single iteration, taking $\Delta t = 1$.

   (*c*) Also solve by the Improved Euler method using repeated iteration, taking $\Delta t = 1$.

   (*d*) Compute the correct values from the analytic formula $q = 2 \times 10^{-6} e^{-0\cdot5t}$.

   (*e*) Draw up a table or sketch a graph showing all four sets of results for comparison.

   (*f*) Find out about Runge Kutta methods of solving ordinary differential equations and use such a method on this (and other) problems.

15. Simulations of chemical, biological and physical systems

INTRODUCTION

There are many problems in real life which are not amenable to solution by existing analytical or numerical techniques. This is particularly so where large systems are studied—e.g. running a chemical plant, flying an aircraft, analysing the ecology of a region. In the first place the system itself is highly complex, and in the second place random or unpredictable occurrences may affect the system. Simulation involves the setting up of an *analogue* or *model* of the system. How the individual elements of the analogue behave is specified and then the analogue is set in motion and its overall behaviour studied. If it mirrors the real-life system then the analogue's details may be changed to see the effect of similar changes in the real-life system. The analogue can be a scale model, or a physically equivalent model, or mathematical equations and logical relations.

As an example of a deterministic system—one governed by exact laws —but one for which an analogue is used for training or testing or to investigate man/machine interactions or to find optimum strategies— we take the problem of landing a spacecraft on a planet or moon (in a very much simplified form).

A DETERMINISTIC SYSTEM: LANDING A SPACECRAFT

Description

A very popular simulation is that of landing a spacecraft on the moon. This is often available on computers and indeed is offered as a program on some of the programmable calculators. In fact the simulation can be done quite easily on any calculator.

Basically the simulation is as follows. The spacecraft is at a certain height (s_0) above the moon (or planet) travelling down at a certain velocity (v_0). There is a limited supply of fuel to burn in the retro-rockets to slow the descent. A successful landing is one which does not need too much fuel and which is 'soft'—i.e. the velocity on impact is low.

Theory

This very brief outline is not essential to running the simulation but may be of interest to those who wish to understand the principles or wish to modify the procedures for themselves.

It is assumed that the intensity of gravity, g, is constant and that each fuel burn is uniform and lasts 1 s, and that each 1 kg of fuel burnt has a retarding effect of 1 ms^{-2}. The height, s, is measured up from the surface and the velocity is measured downwards, i.e. positive velocity for descent and negative for ascent.

The two relevant kinematic equations are

$$s = ut + \tfrac{1}{2}at^2 \quad \text{and} \quad v = u + at$$

After the nth second of the motion the position s_n and velocity v_n are given by

$$s_n = s_{n-1} - v_{n-1} + \tfrac{1}{2}(f-g)$$
$$v_n = v_{n-1} - (f-g)$$

where f is the amount of fuel burned in the nth second and g is the acceleration due to gravity.

Instructions for Running the Simulation

You are taking control of a spacecraft to land it on the Moon or a planet, by firing retro-rockets in 1 s bursts. Each kilogram of fuel burned gives the spacecraft a retardation of 1 m s^{-2}. You cannot burn more than 40 kg in any one s. Choose the destination and note the gravitational pull and suggested fuel for starting from a height of 1000 m, descending at 200 ms^{-1} (Table 15.1).

TABLE 15.1

	Intensity of gravity	Fuel
Moon	1·6 ms^{-2}	240 kg
Earth	9·8 ms^{-2}	330 kg
Mercury	3·7 ms^{-2}	265 kg
Venus	8·5 ms^{-2}	315 kg
Mars	3·8 ms^{-2}	270 kg
Jupiter	26 ms^{-2}	475 kg
Saturn	11 ms^{-2}	340 kg
Uranus	10 ms^{-2}	330 kg
Neptune	12 ms^{-2}	350 kg
Pluto	4·8 ms^{-2}	275 kg

Draw up a table as shown below to indicate the flight details (about 20 s should be enough to land the spacecraft).

Time	Fuel Burn	Fuel Left	Height	Velocity
0	—	240	1000	200
1	. . .	. . .	. . .	. . .

The calculation details are indicated on pages 254 and 255, where one memory is used. The reader can improve the key-stroke sequences if further memories are available.

In each case the particular simulation taken is landing on the Moon ($g \approx 1 \cdot 6$ ms^{-2}) from 1000 m at 200 ms^{-1}. Landing occurs when the height s becomes zero or negative. A soft landing is one for which $v < 2$ ms^{-1}. (see pages 254–255).

TABLE 15.2   *A typical simulation run*

Time	Fuel burn (kg)	Fuel left (kg)	Height (m)	Velocity (ms^{-1})
0	—	240	1000	200
1	10	230	804·2	191·6
2	12	218	617·8	181·2
3	20	198	445·8	162·8
4	20	178	292·2	144·4
5	40	138	167·0	106·0
6	40	98	80·2	67·6
7	30	68	26·8	39·2
8	30	38	1·8	10·8
9	10	28	−18·8	2·4

So a crash landing has occurred here. Once the operator can safely land on the moon using 240 kg fuel he should try to economize on fuel—and it can certainly be done in under 215 kg. Alternatively a faster initial descent can be assumed, or a greater height.

Something to do

Exercise 15.1   Investigate possible strategies for landing the spacecraft to minimize fuel burnt.

SIMULATION OF THE PREPARATION OF ACID AT A GIVEN pH

A flask contains 500 cm^3 of water, and the pH value is checked to be 7·00. A burette contains hydrochloric acid of strength 0·1200 M. The acid is to be added to the water to produce a solution with pH $= 2 \cdot 00$. This simple preparation task is to be used as a test of certain chemistry students' ability, and it is made clear that on no account must the pH fall below 2·00.

(1) *Algebraic—One Memory*

Instructions	Key strokes	Display	Comment	Output Instructions
Input initial velocity	200 STO	200·	v	
(a) Input new fuel burn	10	10	f	Record fuel burn, f, and fuel left
(b) Subtract gravity	− 1·6	1·6	g	
(c) Halve value	÷ 2	2		
(d) Add current height	+ 1000	1000·	s	
(e) Subtract current velocity	− RCL =	804·2	s (new)	Record new height
(f) Enter same fuel burn again	10 +/−	−10·	$-f$	
(g) Add gravity and velocity	+ 1·6 + RCL =	191·6	v (new)	Record new velocity
(h) Store new velocity	STO	191·6	v (new)	
(i) GO BACK TO (a) UNLESS SPACECRAFT HAS LANDED (s (new) ≤ 0)				

[*Note*: 'sum of products' logic machines require line (c) to be: (c) = ÷ 2.]

(2) *Reverse Polish—One Memory*

Instructions	Key strokes	Display	Comment	Output Instructions
Input initial velocity	200 STO	200·	v	
(a) Input new fuel burn	10 enter	10·	f	Record fuel burn, f, and fuel left
(b) Subtract gravity	1·6 −	8·4	$f-g$	
(c) Store in stack	enter enter	8·4	$f-g$	
(d) Halve value	2 ÷	4·2	$\frac{1}{2}(f-g)$	
(e) Add current height	1000 +	1004·2	$s+\frac{1}{2}(f-g)$	
(f) Subtract current velocity	RCL −	804·2	s (new)	Record new height
(g) Clear unwanted value	CLX	0·		
(h) Find new velocity	RCL x⇄y −	191·6	v (new)	Record new velocity
(i) Store new velocity	STO	191·6	v (new)	

(j) GO BACK TO (a) UNLESS SPACECRAFT HAS LANDED (s (new) ≤ 0)

Although pH is really defined in terms of electrode potential an approximate working definition is:

$$pH = -\log[H^+] = -\log\left(\frac{0 \cdot 12V}{V+500}\right) = \log\left(\frac{V+500}{0 \cdot 12V}\right)$$

where V is the volume of acid to be added, in cm^3.

Procedure:

(1) Enter the initial volume V of acid to be added. (A small amount, say up to 5 cm^3, is recommended.)
(2) Calculate the resulting pH.
(3) Add an appropriate extra quantity of acid, trying to avoid pushing the pH below 2·00.
(4) Calculate the new pH. If it is < 2·00 you have failed. If it is 2·00 to 2 d.p. you have satisfactorily completed the assignment. If it is > 2·00 go to (3). (see pages 257–258 for the key-stroke sequences).

General Notes:

(1) The cycles are identical except for the amount of acid added each time.
(2) Extra memories could be used to store the constants in the problem.
(3) Of course other values of acid concentration and water volume and pH required can be used for variety, and differing accuracies demanded.

A Modification:

The simulation can be made more challenging by stipulating a maximum number of separate additions of acid (e.g. 10).

Some things to do

Exercise 15.2   Complete the acid preparation simulation
   (a) with no limit to the additions of acid to the solution permitted.
   (b) limiting yourself to 10 (or less) separate additions of acid.
Exercise 15.3   Re-run the simulation using a different strength acid and a different volume of water.

SIMULATION OF A TITRATION

A 0·100 M solution of aqueous sodium hydroxide is titrated against a 0·100 M solution of hydrochloric acid. 50 cm^3 of the acid is placed in a beaker and the base is titrated in 10 cm^3 at a time until 40 cm^3 has been added, and then in smaller quantities up to 50 cm^3. The pH value is approximately given by

$$pH = -\log[H^+] = -\log\left(\frac{50-V}{50+V}\right)$$

Results Table

	Algebraic key strokes	Display	Comment	Total volume added	pH
Initial set-up	·4 STO	·4	Initial volume, V_1	0·4	
	+ 500 ÷ ·12 RCL =	10425·	$(V+500)/0\cdot12V_1$	0·4	
	log	4·0180761	pH	0·4	4·02
Cycle	5 + RCL = STO	5·4	5 cm³ added, giving V_2	5·4	
	+ 500 ÷ ·12 RCL =	779·93827	$(V_2+500)/0\cdot12V_2$	5·4	
	log	2·8920602	pH	5·4	2·89
Cycle	4 + RCL = STO	9·4	4 cm³ added giving V_3	9·4	
	+ 500 ÷ ·12 RCL =	451·59574	$(V_3+500)/0\cdot12V_3$	9·4	
	log	2·6547498	pH	9·4	2·65
	and so on	. . . . .	. . . . .	. . . . .	. . . . .

(*Algebraic Note:* 'sum of products' logic machines require an [=] between the 500 and the [÷].)

Results Table

	Reverse Polish key strokes	Display	Comment	Total volume added	pH
Initial set-up	500 enter enter	500·	Volume of water		
	·4 STO +	500·4	$V_1 + 500$	0·4	
	·12 RCL ÷	10425·	$(V_1 + 500)/0·12V_1$	0·4	
	log	4·0180761	pH	0·4	4·02
Cycle	CLX 5 RCL + STO	5·4	V_2	5·4	
	+ ·12 ÷ RCL ÷	779·93827	$(V_2 + 500)/0·12V_2$	5·4	
	log	2·8920602	pH	5·4	2·89
Cycle	CLX 4 RCL + STO	9·4	V_3	9·4	
	+ ·12 ÷ RCL ÷	451·59574	$(V_3 + 500)/0·12V_3$	9·4	
	log	2·6547498	pH	9·4	2·65
and so on		· · · · ·	· · · · ·	· · · · ·	

(*Reverse Polish Note*: The 500 (Volume of water) is stored in the stack using the facility that when the stack drops it regenerates what was in the top register ('T').)

where V is the quantity of acid now in the solution, provided there is an excess of base.

Something to do
Exercise 15.4

(*a*) Calculate the pH values obtained working very gradually up to $V = 50$, and complete Table A.

TABLE A		TABLE B	
Total NaOH added V cm³	pH value	Total NaOH added V cm³	pH value
0		50·00001	
10		50·0001	
20		50·001	
30		50·01	
40		50·1	
45		50·5	
49		51	
49·5		55	
49·9		75	
49·99		100	
49·999		1000	
49·9999		10000	
49·99999		100000	

(*b*) If too much base is now added for neutralization, the appropriate formula is

$$pH = 14 - \log \left(\frac{V + 50}{V - 50} \right)$$

Complete Table B using this formula.

SIMULATION INVOLVING A BUFFER SOLUTION

Buffer solutions are used to reduce the otherwise excessively large changes in pH which occur when a concentrated acid or base is added to a solution. V cm³ of a 1 M solution of hydrochloric acid are added to 500 cm³ of a buffer solution consisting of 1 mol benzoic acid per l and 1 mol sodium benzoate per l. The system is described by:

$$[H^+] = K \frac{[C_6H_5COOH]}{[C_6H_5COO^-]}$$

$$= 6·5 \times 10^{-5} \times \frac{(500 + V)/1000}{(500 - V)/1000}$$

The pH can be approximately defined as

$$pH = -\log [H^+] = -\log [6\cdot5 \times 10^{-5} \times (500 + V)/(500 - V)]$$
$$= 4\cdot19 + \log [(500 - V)/(500 + V)]$$

Something to do

Exercise 15.5   Devise a calculator procedure to simulate the addition of the acid to the buffer solution.

Some interesting chemical simulations for pocket calculators have been described by D. K. Holdsworth (1976), A. Foglio Para and E. Lazzarini (1974), R. B. Snadden and O. Runquist (1975) and M. J. Perry (1977).

PROBABILISTIC SIMULATIONS

If there are random (unpredictable) occurrences of importance in a system—e.g. pipe fractures and other delays in chemical plants, deaths in animal populations, queues of callers in a telephone exchange—then the simulation analogue must reflect this random behaviour. This requires the use of 'random numbers'. Another use for random numbers is where a deterministic problem exists which is too difficult or indeed impossible to solve analytically—then a probabilistic model is made which can be treated experimentally (this is often known as the Monte Carlo technique).

Tables of random numbers are often used but with the availability of computers researchers sometimes generate their own as they are needed. Although random numbers can be generated using a calculator the need for large numbers of them makes the use of prepared random number tables attractive for large scale problems. For small problems the calculator may be quite satisfactory but then so are dice!

What are random numbers?

A distinction must be made between *true random numbers* and *pseudo-random numbers*. True random numbers can only be generated by some physical process: throwing a die, tossing a coin, playing roulette, counting α-particle emissions, measuring noise in electrical systems (e.g. ERNIE) etc. In such cases although we know the appropriate 'law of large numbers' there is no known deterministic law to enable us to predict *individual* events. These are—or appear to be—truly random. On the other hand pseudo-random numbers are in effect members of cycles of numbers produced mathematically, which share many common properties with true random numbers but which are really deterministic and reproducible. Schemes can be found to generate pseudo-random numbers for any given problem. In many cases pseudo-random numbers are preferable in simulations because then a re-run can be made with perhaps one part of the system varied, and 'nature'—represented by the random numbers—can be held constant.

Congruence Methods

The methods of generating pseudo-random numbers most often used in practice are multiplicative congruence methods. One such method is based on the recurrence relation $y_n = Ky_{n-1}$ (mod A) and the scaling equation $x_n = y_n/A$.

K, A, and the $\{y_i\}$ are integers. The $\{x_i\}$ are in the range $[0, 1]$. The expression Ky_{n-1} (mod A) means that we take the remainder when the product Ky_{n-1} is divided by A: e.g. 8×3 (mod 9) is 6. The recurrence relation produces a sequence $\{y_i\}$ of pseudo-random integers in the range $[1, A-1]$ and the scaling equation normalizes the range to $[0, 1]$. The choice of parameters K, A and the starting value y_0 are governed by certain rules depending on Number Theory which we shall not go into here. Usually A is taken to be a power of 2; K is taken to be of the form $8t \pm 3$ where t is an integer chosen such that $K \approx \sqrt{A}$; y_0 is any odd integer in the range $1 \leqslant y_0 < A$. This ensures that the maximum cycle length of $A/4$ is attained.

Some programmable calculator handbooks have procedures similar to this for generating pseudo-random numbers.

Example

A realistic sequence is

$$A = 2^{16} = 65536$$

$$K = 2^8 + 3 = 259$$

$$y_0 = 725 \text{ (say)}$$

$$y_1 = 259 \times 725 \text{ (mod } A) = 187775 \text{ (mod } A).$$

Now we generally will want the normalized value x_1 which is y_1/A. In this case it is easier to find x_1 first because $18775 \div A$ will give as its fractional part the x_1 we seek:

$$187775/65536 = 2 \cdot 8652191$$

so $x_1 = 0 \cdot 8652191$.

We round *down* to 3 d.p. to get $0 \cdot 865$ as our random number to ensure a uniform distribution. Now to get y_1 we multiply x_1 by A. The key-stroke sequence is illustrated on page 262.

	Algebraic key strokes	Display	Comment	Reverse Polish key strokes	
(a)	2 [y^x] 16 [=] [STO]	65536·	A	2 [enter] 16 [y^x] [STO]	(a)
(b)	259 [×] 725	725·	y_0	259 [enter] [enter] [enter]	(b)
(c)	[÷] [RCL] [=]	2·8652191	Ky_0/A	[×] [RCL] [÷]	(c)
(d)	[−] 2 [=]	·8652191	x_1	2 [−]	(d)
(e)	[×] [RCL] [=]	56703·	y_1 (may need rounding off)	[RCL] [×] [RCL] [÷]	(e)
Cycle { (f)	[×] 259 [÷] [RCL] [=]	224·09175	Ky_1/A	224 [−]	(f) } Cycle
(g)	[−] 224 [=]	·091751	x_2	[RCL] [×]	(g)
(h)	[×] [RCL] [=]	6013·	y_2 (may need rounding off)		(h)
	and so on	. . .	. . .	and so on	

Note: The Reverse Polish line (*b*) 259 [enter] [enter] [enter] fills up the stack with K which is repeatedly needed. On line (*f*) the [×] uses K from the stack. On some machines the K is automatically reproduced in the T register so the stack is kept 'topped up'. If this does not happen the method will not work and 259 must be repeatedly inserted (or stored). In that case line (*f*) could be 259 [×] [RCL] [÷] .

The resulting sequence (of 16384 numbers) will be uniformly spread from 0 to 1. Thus if 3 d.p. rounded down numbers are taken they should all occur more or less equally frequently (about sixteen of each from ·000 to ·999). Therefore we have a source of 3-digit numbers which can then be used to give numbers of any length.

For example the two 3-digit numbers ·129 and then ·726 could be used to give:

Six 1-digit numbers 1, 2, 9, 7, 2, 6
or Three 2-digit numbers 12, 97, 26
or One 6-digit number 129726.

BROWNIAN MOTION SIMULATION

Assume that a particle begins at the centre of a 2-dimensional grid. It can move in any of eight directions on the grid (at 45° intervals). Generate a random digit and plot the path as follows:

   (i) If the digit is 0 the particle stays where it is.

   (ii) If the digit is 1 to 8 the particle moves as shown in fig. 15.1.

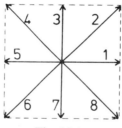

Fig. 15.1

   (iii) If the digit is 9 the previous motion is repeated.

For example the sequence of calculator-generated digits:

2, 3, 8, 2, 5, 8, 9, 1, 4, 6, 9

would produce the simulated Brownian motion shown in fig. 15.2.

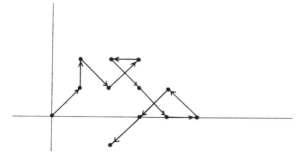

Fig. 15.2   Simulation of Brownian motion.

An alternative scheme is to use only digits 1 to 6 to represent movements (in 2 or 3 dimensions). A more realistic picture may be obtained by joining up every fourth dot so that a variety of different path lengths (one per 4 time units) is displayed.

ENTROPY LEVEL SIMULATION

If an evacuated cylinder is connected to a cylinder containing an ideal gas, what will happen? The Second Law of Thermodynamics predicts that the gas will rush into the evacuated flask until the pressures equalize and then the system is stabilized. (The *entropy* rises to a maximum.)

A very simple model of this was provided (in 1907) by P. and T. Ehrenfest. Jar A contains n balls, marked 1 to n, and Jar B is empty. A hat contains discs on which are marked, in equal proportions, the digits $1, 2, \ldots n$. Every time a disc is drawn from the hat the corresponding ball is moved from whichever jar it is in to the other jar. Initially, of course, the predominant trend is from Jar A to Jar B but movements the other way can take place.

Taking the rather small but manageable case of 10 balls (marked 0 to 9 for convenience rather than 1 to 10) we simulate the molecular behaviour as follows. We generate a stream of 1-digit random numbers. We keep a record of the positions of the individual (imaginary) balls and keep a tally of the numbers of balls found in each jar.

Stream of digits: 6, 8, 7, 3, 6; 1, 7, 1, 5, 2; 4, 5, 7, 4, 9; 2, 6, 9, 0, 1; 6, 2, 2, 1, 4.

Time	Jar A	Jar B	Number in Jar A
Initially	0, 1, 2, 3, 4, 5, 6, 7, 8, 9	—	10
After 5 changes	0, 1, 2, 4, 5, 6, 9	3, 7, 8	7
After 10 changes	0, 1, 4, 6, 7, 9	2, 3, 5, 8	6
After 15 changes	0, 1, 4, 5, 6	2, 3, 7, 8, 9	5
After 20 changes	2, 4, 5, 9	0, 1, 3, 6, 7, 8	4
After 25 changes	1, 2, 5, 6, 9	0, 3, 4, 7, 8	5

The simulation does indicate the rise in entropy (levelling out). A model using more (imaginary) balls would give a better picture.

Some things to do

Use a calculator to generate pseudo-random numbers, and hence:

Exercise 15.6   Simulate 2-dimensional Brownian motion.

Exercise 15.7   Simulate the Ehrenfest model for 25 balls. This can be done either by generating 2-digit numbers 0 to 99, dividing by 4 and rounding down to give random numbers from 0 to 24, or by generating the required numbers directly.

Some interesting problems in population dynamics have been posed by A. Engel (1973). We now look at some mathematical models based on his suggestions. The first is a very simple model used to introduce two much more realistic and complex models.

A simple model: small creatures

For a population of small creatures $\frac{5}{16}$ die in their first year of life (i.e. aged 0) and none live to reach 2 years of age (i.e. the rest die aged 1). The average number of *female* offspring produced *per female aged 0* is $\frac{3}{5}$ and *per female aged 1* is $\frac{4}{5}$. The creatures breed once per year.

Let x and y be the populations of females aged 0 and 1 respectively.
(a) What will the population be in future years?
(b) What will the stable growth rate be after a few years?
(c) What will be the stable proportions of 0 year olds and 1 year olds in the total population?

To answer these questions we must set up a mathematical model. There are two equations which describe the system:

$$0 \text{ year olds:} \quad x_{n+1} \quad = \quad \tfrac{3}{5}x_n \quad + \quad \tfrac{4}{5}y_n \ldots \text{I}$$

New 0 year olds   Offspring from 0 year olds   Offspring from 1 year olds

$$1 \text{ year olds:} \quad y_{n+1} \quad = \quad \tfrac{11}{16}x_n \ldots \ldots \ldots \text{II}$$

New 1 year olds   Surviving 0 year olds

Writing in decimal form: $x_{n+1} = 0{\cdot}6x_n + 0{\cdot}8y_n \ldots \text{I}$

$$y_{n+1} = 0{\cdot}6875x_n \ldots \ldots \text{II}$$

Starting with $x_0 = 5000$ and $y_0 = 4000$ we can use the calculator to find the future population levels: x_1, y_1, then $x_2, y_2, \ldots$

$x_1 = 0{\cdot}6x_0 + 0{\cdot}8y_0$

$\therefore \quad x_1 = 0{\cdot}6 \times 5000 + 0{\cdot}8 \times 4000 = 6200$

$y_1 = 0{\cdot}6875x_0$

$\therefore \quad y_1 = 0{\cdot}6875 \times 5000 \approx 3438$ (rounding off to nearest integer)

$x_2 = 0{\cdot}6x_1 + 0{\cdot}8y_1 = 0{\cdot}6 \times 6200 + 0{\cdot}8 \times 3438 \approx 6470$

$y_2 = 0{\cdot}6875x_1 = 0{\cdot}6875 \times 6200 \approx 4263$

Some further values are included in Table 15.3.

TABLE 15.3

n	0	1	2	3	4	5
x_n	5000	6200	6470	7292	7934	8771
y_n	4000	3438	4263	4448	5013	5455

It is quite possible to devise procedures for calculators such that no re-entry of the x_n and y_n values is necessary. (This problem is closely related to that of the Foxes and Rabbits population calculations of Chapter 12, page 209.) The reader is invited to devise such a procedure for his own calculator—it is not so simple as it at first appears if economy of key strokes is sought!

Something to do

Exercise 15.8   The answers to the three questions posed above can all be given using the results of Table 15.3 (which the reader should extend to $n=10$ or more):
   (a) adding the two columns will give the 'total population' x_n+y_n for any year.
   (b) the stable growth rate will be found by calculating the values

$$\frac{\text{Total population at } n}{\text{Total population at } n-1} \quad \text{i.e.} \quad \frac{x_n+y_n}{x_{n-1}+y_{n-1}}$$

until the sequence settles down to a fixed value.
   (c) The ratio $\{x_n/(x_n+y_n)\} \times 100$ will give the percentage of 0 year olds in the population and calculating successive values of this should show it settling down to a constant value.
   All your results can be set out in one table:

Year n	Females aged 0 x_n	Females aged 1 y_n	Total female population x_n+y_n	% females aged 0 $\dfrac{100x_n}{x_n+y_n}$	Growth rate $\dfrac{x_n+y_n}{x_{n-1}+y_{n-1}}$
0	5000	4000	9000	55·56	—
1	6200	3438	9638	64·33	1·071
2	6470	4263	10733	60·28	1·114
...	....	....	.....	.....	.....

A Deterministic model: human population

A more realistic model, for the *human female* population of a country will involve more details for the population structure—e.g. numbers of females of various ages and their associated death rates.

The following approximate data (Table 15.4) are based on the United Kingdom female population for 1974. [*Source:* Annual Abstract of Statistics, 1975, Central Statistical Office, H.M.S.O. London.]

TABLE 15.4

	Age group (years)	Numbers (thousands)
A	0–9	4242
B	10–19	4175
C	20–29	3968
D	30–39	3270
E	40–49	3266
F	50 and over	9828

If we deal with time units of 10 years then the above represents the situation at time 0. If we know the *death rates* for each group over the period we can forecast the new group sizes for B to F in 10 years time. And given the *birth rates* we can forecast the size of group A. Some estimated figures are given in Table 15.5.

TABLE 15.5

Age group (years)	Female death rate % for 10 year period	Female birth rate % for 10 year period
A 0–9	2·00	0
B 10–19	0·29	11
C 20–29	0·49	77
D 30–39	1·00	22
E 40–49	2·68	2
F 50 and over	27·64	0

If we let the variables a, b, c, d, e, f represent the numbers in the various age categories we get the following system of equations to give 10-year forecasts, from period n to period $n+1$:

(i) $a_{n+1} = 0·11b_n + 0·77c_n + 0·22d_n + 0·02e_n$ (Births from the various contributing age groups)

(ii) $b_{n+1} = 0·98a_n$ (98% of those aged 0–9 survive 10 years to be in the 10–19 age group)

(iii) $c_{n+1} = 0·9971b_n$

(iv) $d_{n+1} = 0·9951c_n$

(v) $e_{n+1} = 0·99d_n$

(vi) $f_{n+1} = 0·9732e_n + 0·7236f_n$ (97·32% of the 40–49 age group survive 10 years and so do 72·36% of the 50+ age group).

The population dynamics can be easily computed from these figures using a calculator and the results are best set out as in Table 15.6.

TABLE 15.6

Period, n	0–9 a_n	10–19 b_n	20–29 c_n	30–39 d_n	40–49 e_n	50+ f_n
0	4242	4175	3968	3270	3266	9828
1	4299	4157	4163	3949	3237	10290
2	...	...	...	...	...	...

As an example of the calculations involved:

$$a_1 = 0.11b_0 + 0.77c_0 + 0.22d_0 + 0.02e_0$$
$$= 0.11 \times 4175 + 0.77 \times 3968 + 0.22 \times 3270 + 0.02 \times 3266$$
$$\approx 4299$$
$$b_1 = 0.98a_0$$
$$= 0.98 \times 4242$$
$$\approx 4157$$

and so on.

Each line of the table is produced from the previous line.

Something to do

Exercise 15.9   Run the population model through 10 periods or so. Does there appear to be any pattern emerging?

A Probabilistic model: human population

The previous population dynamics model is deterministic—everything is fixed and no unpredictable ('random') behaviour is permitted. Simulation of the real-life situation should include some unpredictable variation—in the reproductive rates particularly—giving a probabilistic model. This can be quite easily done by altering the birth rates (say) from period to period depending on the throw of a die:
Procedure:

(1) Throw a six-sided die and if
    (*a*) 1 turns up: leave all rates as they were for the last period
    (*b*) 2 turns up: increase birth rates by 5% (of their values for the last period)
    (*c*) 3 turns up: decrease birth rates by 5%
    (*d*) 4 turns up: increase birth rates by 10%
    (*e*) 5 turns up: decrease birth rates by 10%
    (*f*) 6 turns up: reduce 50+ death rate by 5%.

Note: round off all rates to 3 s.f.

(2) Run one cycle with the new rates, rounding off all population sizes to the nearest integer.

(3) Go back to (1).

Some things to do

Exercise 15.10   Run the probabilistic model through 10 periods or so and compare with the deterministic model. Is there any significant difference?

Exercise 15.11   Devise some alternative changes to the model to be decided 'randomly' on the throw of a die.

Exercise 15.12   *An open-ended problem: Drosophila melanogaster*   The normal length of life of this fruit-fly is about 28 days when kept at 25°C: egg stage 1 day; larva 3–11 days; pupa 2–8 days, adult about 14 days. The female may lay 100 eggs. Devise a mathematical model incorporating these figures to simulate the changes in population structure and size with time of a population of D. melanogaster. [*Source:* Biological Data Book, 2nd Edition, Vols I, II. Edited by P. L. Altman & D. S. Dittmer, Fed. Amer. Socs. Exper. Biol.]

16. Programmable calculators—the next logical step

It is fitting that this last chapter should be devoted to the programmable calculator. These are not, as some may think, simply the next generation of electronic calculator but rather a new species which owe their parentage jointly to the calculator and to the computer. The major difference between programmable calculators and computers is in the amount of storage space for programs which is available and the fact that calculators, so far, do not accept an alpha-numeric input. Computers are quite happy to accept messages made up of letters and words as well as numbers, in fact most high-level computer languages are quite literal, but not so the calculator. It has to have its entries as numbers and special key depressions. Some machines do have a capacity for dealing with English words, the Texas Instruments SR 60 being one (fig. 16.1).

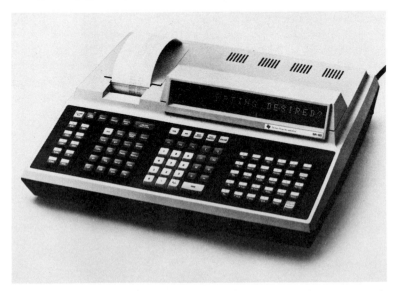

Fig. 16.1   The Texas Instruments SR 60 which has an alpha-numeric display.

The words, though, do not affect the program in any way, they are merely entered by the operator, stored and then regurgitated at the point in the program designated by the operator. The most useful feature of incorporating words into a program in this fashion is that they can act as prompts for an unskilled operator to tell him what should be done next. However we have jumped ahead of our story and should go back a bit. The first programmable machines were simply electronic calculators whose memory capabilities had been extended so that a series of *key strokes* rather than just numbers could be remembered. With these the operator solved a simple problem by pressing the keys in the necessary sequence and leaving 'spaces' where data needed to be entered. A typical example might be to evaluate

$$\sqrt{(A^2 + B^2)}$$

The key sequence which would be used might be

| Start | | Halt | | x² | | STO | | Halt | | x² | | + | | RCL | | = | | √x | | Stop |

The Start command tells the calculator that it has to remember the key strokes which follow; the Halt command means that at this point during the execution of the program the calculator will halt for data to be entered, via the keyboard, while the Stop command lets the calculator know that the writing of the program has finished and that it should now go back to the start of the program and be ready to run it.

The user can now use this program to calculate $\sqrt{(A^2 + B^2)}$ for any pair of numbers he likes. Suppose he chooses $A = 3$ and $B = 5$. He will enter 3 and follow this entry by depressing the RUN key. The display will blank out for a few seconds with perhaps a hyphen appearing to show that calculation is in process, then the hyphen will disappear indicating that the next number is to be entered. (It is possible that the intermediate result, 9 in this case, would be shown.) 5 is now entered by the operator, followed by depressing the RUN key and after a couple of seconds the display shows 5·8309519 which is the answer for that particular pair of numbers. If required other pairs of numbers can be entered and the results calculated in the same manner.

It will be seen that this type of programmable calculator *stores the key strokes*, thereby relieving the user of the drudgery necessary in entering the key strokes every time should a particular problem need working out for several sets of data. The Control ROM is acting under the command of the stored key strokes just as if the keys were being depressed by the user at that time.

This programming facility has now been extended so that more recent machines not only store key strokes but can also *act on logical commands* (e.g. 'if ... then ...') incorporated into the program sequence. Another extension is that the program can be transferred out of the calculator, being stored on a small magnetic card for use at a later date.

In some of the newer programmable machines it is *key phrases* rather than *key strokes* which are stored. A key phrase is a sequence of key strokes that together perform a single operation—in some cases three separate key strokes are merged to form one key phrase and this fact should be remembered when comparing the sizes of program storage for different makes. Note that the 'function' or 'shift' key which selects the second feature of a key does not normally count as a key stroke.

It is best not to have to key in numbers into a program since for each *digit* entered one key-stroke store is used—it is much better to store the number in a memory before running the program and recall it from the memory via a key stroke. It is key-stroke recall which makes the programmable calculator so powerful since any of the algorithms provided on the calculator is available at the expense of just a single program step.

When one looks at a programmable calculator it will be noticed that there are some additional keys which are not found on the other types of calculator and we therefore need to define their purpose:

| LRN | or | W Prgm | *Learn* or *write program*. This switch puts the calculator into the correct mode so that the subsequent key strokes are stored in the program memory. The display format will also change so that the key code for the key depressed is shown together with a number showing which memory location is being used. The key code usually refers to the particular row and column the key is located in. The Casio programmable calculator is interesting in that it displays the codes and locations for the two most recent operations as well as for the current one.

| HLT | or | R/S | *Halt* or *Stop*. If this command is encountered during a program, control is returned to the keyboard. It is usually used when data for a particular program needs to be entered before the program can proceed. On some machines it signifies the end of the program.

| READ | *Read in* the program which is stored on a magnetic card into the calculator's own internal memory. The contents of the card are not destroyed and may be reused.

[A] [B] [C] etc.

These are what are known as *user-definable* keys, and may number up to ten separate keys. They are conveniently used as labels for particular programs which have already been written by the user, and which are stored in the calculator. By pressing the appropriate key that particular program will be executed. Normally they are used to follow a data input and so to restart the calculation. Alternatively they may be used as a new function key to provide a function which is not part of the keyboard instruction set.

[LBL]

Label. Using this key, often in conjunction with others, a particular point in a program can be labelled for future reference in the program.

[GTO]

The *Go To* command tells the calculator executing the program to jump to a particular program location. This location may be referenced by its location number (step number) within the program, e.g. [GTO] [057] immediately re-positions the program pointer to location 57 so the instruction stored there will be the next executed. Alternatively the position may be defined by a previously used label command. For example [GTO] [A] will position the program pointer to the step which has previously been identified by using [LBL] [A] keying. With the [GTO] instruction the program continues sequentially from the new location.

[GSB] or [SBR]

These are used to go into a *subroutine* sequence which may be labelled in several different ways, by user-definable keys or step numbers for example.

[RTN]

This is the *return* key which is the last command of any subroutine since it returns the operation back to the main program.

if pos [x < y]
[x = 0] etc.

There is a wide variety of these *conditional branching instructions* which greatly add to the versatility of the programmable machine since during execution of the program the calculator can make decisions based on the results of testing data. These commands are always followed by an implied Go To instruction. This is because if the test is true the next program step will be executed; if the data test is false then the program skips one step and continues sequentially.

SF 1   TF 1
st flg   if flg

These are the so called *Flags*, which act as data route indicators or as conditional commands within a program.

DSZ

Decrement and skip on zero. A number is stored in one of the user-addressable memories while an iterative loop subroutine is being run. Each time the loop goes round one is subtracted from the stored number which is then compared with zero. If it is not zero the program continues, if zero the program will skip to the next specified program address.

SST   BST

Single step and *back step* are used for editing and debugging programs. SST executes the program one step at a time allowing the user to view the codes for the various stored instructions. BST causes the program counter to be decremented by one so that entries made previously may be viewed.

DEL

Operation of this key will *delete* the displayed instruction from the program memory and all subsequent instructions move up one step.

INS

A new instruction can be *inserted* between two previously entered instructions by using this key.

PAUSE

Should an intermediate result need to be seen during the execution of a program this key will allow the program to *stop temporarily* at that point for about $\frac{1}{2}$ to 1 s and display the number before it continues. It may also be used if it is necessary to key in data or load a new magnetic program card during the course of a program.

The above list of the major instructions which are available on programmable calculators should give an indication as to how powerful they are in terms of computing capability. Those familiar with computer programming languages such as FORTRAN, ALGOL or BASIC will have recognized many common features.

When writing a program for one of these machines many of the attributes of a computer programmer are needed—not least of which is logical thinking. The best way to start, once the problem has been defined, is to draw a flow diagram of the various steps which are needed to solve it. Once this is done it is a relatively simple matter to convert it into the necessary key strokes which are usually written onto a special coding form supplied by the manufacturers.

As an example, we have taken a sample program produced by Texas Instruments for use on their SR 52. It is assumed that you wish to calculate your bank charges knowing that

£0.10 is charged per cheque for the first five drawn in a month
£0.09 is charged per cheque for the next five drawn in a month
£0.08 is charged per cheque for the next five drawn in a month
£0.07 is charged for each further cheque drawn in any month.

Obviously we shall need to enter the number of cheques drawn in that month and then carry out some conditional tests, e.g. 'Is the number bigger than 5?' and so on. The flow diagram which results is shown in fig. 16.2.

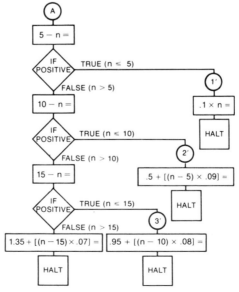

Fig. 16.2 Flow diagram to solve the bank charges problem outlined in the text. n represents the number of cheques drawn in any one month.

The different shaped boxes are used to identify the different processes which are being carried out. 'A' is the user-defined key which will heve to be pressed to commence the program. One can follow the various logical steps which must be taken to work out the charge for different numbers of cheques. This is now translated into key strokes which are shown in fig. 16.3.

The following points are worth noting. At steps 002, 003, 004 the number entered is stored in memory 19. The user-addressable key 'E' has been used to define this memory location, so each time 'E' is encountered the contents of that memory are read—this is done to save storage locations. The subroutine which does this is shown in fig. 16.3h.

Once entered the program is then run through to check and edit it. There is little point in trying to optimize the number of steps needed for

Fig. 16.3   The actual key strokes which are needed to implement the 'bank charges' program on an SR 52. The LOC column refers to the position which the key stroke occupies within the program memory, the CODE is the number appearing in the display to indicate to the user which key has been pressed. The various parts of this program can be identified by referring to the flow diagram shown in fig. 16.2. (Example by courtesy of Texas Instruments.)

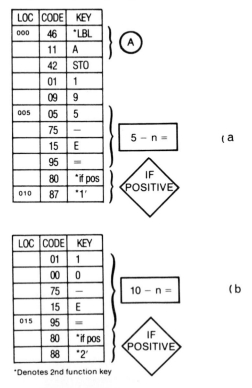

LOC	CODE	KEY
000	46	*LBL
	11	A
	42	STO
	01	1
	09	9
005	05	5
	75	−
	15	E
	95	=
	80	*if pos
010	87	*1'

(a $5 - n =$ IF POSITIVE

LOC	CODE	KEY
	01	1
	00	0
	75	−
	15	E
015	95	=
	80	*if pos
	88	*2'

(b $10 - n =$ IF POSITIVE

*Denotes 2nd function key

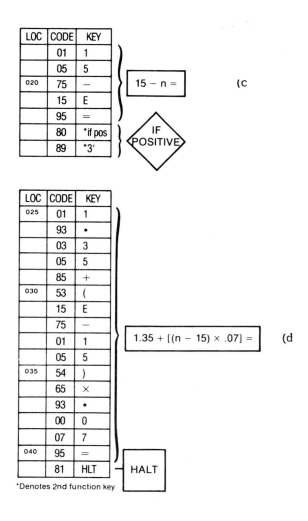

LOC	CODE	KEY
	01	1
	05	5
020	75	—
	15	E
	95	=
	80	*if pos
	89	*3′

15 − n =   (c

IF POSITIVE

LOC	CODE	KEY
025	01	1
	93	•
	03	3
	05	5
	85	+
030	53	(
	15	E
	75	—
	01	1
	05	5
035	54	)
	65	×
	93	•
	00	0
	07	7
040	95	=
	81	HLT

1.35 + [(n − 15) × .07] =   (d

HALT

*Denotes 2nd function key

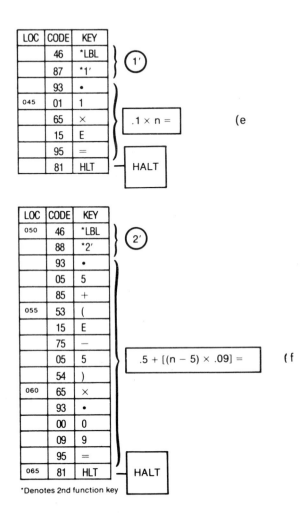

LOC	CODE	KEY
	46	*LBL
	87	*1'
	93	•
045	01	1
	65	×
	15	E
	95	=
	81	HLT

$\}$ 1'

.1 × n =   (e

HALT

LOC	CODE	KEY
050	46	*LBL
	88	*2'
	93	•
	05	5
	85	+
055	53	(
	15	E
	75	−
	05	5
	54	)
060	65	×
	93	•
	00	0
	09	9
	95	=
065	81	HLT

$\}$ 2'

.5 + [(n − 5) × .09] =   (f

HALT

*Denotes 2nd function key

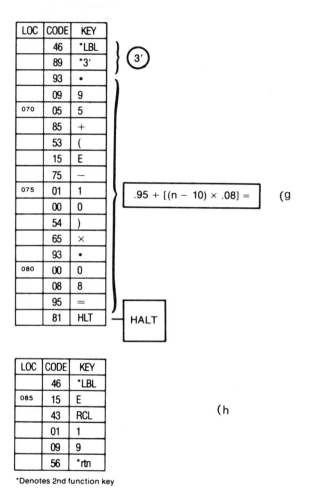

LOC	CODE	KEY
	46	*LBL
	89	*3′
	93	•
	09	9
070	05	5
	85	+
	53	(
	15	E
	75	—
075	01	1
	00	0
	54	)
	65	×
	93	•
080	00	0
	08	8
	95	=
	81	HLT

(3′)

$.95 + [(n - 10) \times .08] =$ (g

HALT

LOC	CODE	KEY
	46	*LBL
085	15	E
	43	RCL
	01	1
	09	9
	56	*rtn

(h

*Denotes 2nd function key

a particular program unless of course the program does not fit into the available storage space. Most of the time spent in trying to condense a program is wasted except for the personal satisfaction which it may bring.

Obviously this is quite a simple example but it does show how easy it is to program a machine of this type.

As more and more people realize this it is likely that they will turn to the programmable calculator—or its offspring—rather than use expensive computer time. In fact, linked to a printer, which the latest Texas Instruments and Hewlett Packard ones can be, they should become a standard piece of office or laboratory equipment.

The two classes of programmable machines now available are firstly those which have an internal program memory and secondly those where a program may be retained externally on a magnetic card as for the Hewlett Packard and Texas Instruments machines or on magnetic tape in the case of the Rockwell Sumlock range, or in special semiconductor file cartridges as used in the Novus 7100.

The first category tend to have a program capacity of between fifty and a hundred program steps with about ten user-addressable memories. In addition to the programming facility they will have various scientific, mathematical or financial functions available on the keyboard. Usually they are handheld and have an LED display though it may be possible to plug them into a separate printer unit. The early machines in this class 'lost' the program when the power was switched off but the latest models incorporate small batteries so that the IC's remain activated even though the calculator is turned off. Casio and Hewlett Packard provide this added feature.

The magnetic card machines have upwards of a hundred program steps with some twenty memories. It is even possible with the larger desk top machines to purchase expansion modules which further increase the program and memory capabilities. The magnetic cards are about 10 mm × 50 mm and are propelled into the calculator by a tiny motor in order that the information on them may be read. This is not apparently as easy as it may sound since the designers had to cope with many problems ranging from motor speed control to dirty cards. The heads which read the cards have to be aligned within a quarter of a degree. Hewlett Packard use a twin track recording technique on the cards used for the HP 65. Here two tracks of varying magnetic flux are placed side by side—one track recording the binary '1's and the other the binary '0's as they occur in the data stream. The '1' track will contain a flux reversal for each '1' but no flux change for a '0'. This method is self clocking and maximizes the system's tolerance to head misalignment and motor speed variations. It is probable that other card readers use a two-track system but they may use one track for the clock pulses whilst the other records all the data. The cards are provided with a small cut-out which prevents the recorded program from being over-written. Cards are not transferable between makes, though the desk top HP 97 uses the same cards as the handheld HP 67 with the result that programs may be transferred between them (fig. 16.4).

The latest development has been the introduction by Texas Instruments of 'solid state software' modules which plug into their TI 58 and TI 59 programmable calculators. Each of these modules contains up to 5000 program steps and allows the user to tailor his calculator to solve problems within his own subject area. The calculator is initially fitted with a 'master library' module which contains 25 programs

Fig. 16.4 The Hewlett Packard HP 67 and HP 97 programmable calculators showing the small magnetic cards used to store their programs.

ranging from solution of linear simultaneous equations to unit conversions. The programs are called up from the keyboard and may be used independently or as part of a longer program. Other modules cover such specialized topics as surveying and navigation. Compatible with these two calculators is a print cradle into which they fit and which will not only give a hard copy of all calculations but will also plot graphs and print instructions.

It is not the intention of this chapter to teach you how to program since this is done by the manufacturers in their handbooks. Rather we would draw your attention to these powerful calculators in the hope that by becoming aware of them some of your calculating problems may be eased. Many of the mathematical examples in this book require a large number of key strokes to solve them—if you have to solve the same expression a number of times then a lot of effort is required not only to enter it but to ensure that it is entered without errors. With a programmable machine—especially one with external storage—the program once written and debugged is ready for instant and repeated use.

We have tended to think in terms of mathematical or scientific uses when describing examples of programs for the programmable calculator but in fact their use is much, much wider. An acquaintance 'borrows' the office SR 52 at weekends to take sailing with him since he has a set of prerecorded programs which he uses for the various navigational problems he needs to solve. In fact using a programmable calculator in this way means that sights of the sun or stars are literally worked out with a few key presses taking a minute or so compared with the fifteen minutes required using tables and long arithmetic. Before long one might see the central heating salesman who, needing to compute the total heat capacity of the boiler for a given installation, the size of radiators for each room and to give an estimate of cost, doing so simply and accurately on a programmable calculator. It should be possible too for the insurance salesman to have suitably preprogrammed cards containing all the relevant data he requires to give quotations for endowment policies etc.

Anybody having a lot of repetitive calculations to perform or who is using a computer for non-sophisticated problems should seriously consider using a programmable calculator.

Appendix
The games calculators play

The serious part of the book is now behind us and we shall dwell for a few moments looking at another way in which the calculator can be utilized. In addition to being a very powerful arithmetical tool the calculator can also be used to play games—and here are a few ideas.

1. Most people will have found out that if a calculator is turned round then the upside down figures in the display can look like certain letters of the alphabet—more so if you have a good imagination. The list below gives the figures and their corresponding letters:

$$0 = O \qquad 5 = S$$
$$1 = I \qquad 6 = g$$
$$2 = Z \qquad 7 = L$$
$$3 = E \qquad 8 = B$$
$$4 = h \qquad 9 = G$$

These can be used to write various messages; a typical example might be 7100553, which when turned around reads ESSO OIL.

We reproduce with permission a crossword from *New Electronics* which has conventional clues and answers but unconventional supplied solutions (p. 297). The answers are checked by performing the given calculation—as a normal 'chain calculation' without the operators being given any hierarchical significance—and then turning the calculator upside down to read the answer as a word rather than as a number.

Clues

ACROSS

1 Sounds like a killing means of transport (6)
4 Get too involved in jobs essential in part (6)
7 It's a laugh to break a leg in 26 down (6)
9 Restrict vision to a tenant (6)
12 Nothing sounds like a violin player following a woodwind instrument (4)
14 What the walker wears to send away I hear! (4)
16 An insect returns from the ebb (3)
17 He should account for this (4)
19 Could be a fast berry but it sounds slow (4)
22 Overseers of Studs, maybe (6)
25 Fumble for a Guinea to give to the Scarecrow (6)
27 His French produces debits with Spanish articles (6)
28 Stare at a spectacle (6)

DOWN

1 Sounds as if he's largely depressed (5)
2 Produce from a coop, and urge on! (4)
3 Lower this implement to dance (3)
4 Well well, a lucky strike! (3)
5 Alternatively in shelters (4)
6 An explosive covering for a crustacean (5)
8 For the Learner given this honour, an ear-piercing part (4)
10 Put in a Pound to see these slippery creatures (4)
11 Partly probes Eiger—to surround and starve (7)
13 Be cooked with lubrication, maybe? (4)
15 A Spanish wolf returns to find part of a Greek penny (4)
18 Mislay a flatfish (4)
19 Can you hear this store taking a deep breath? (4)
20 It's not slander that I am ringing (5)
21 Fat sort of outsize insect (5)
23 Relatively short! (3)
24 Ask for help, initially (3)
25 Plead for a new beginning (3)
26 I join a horse in a carriage (3)

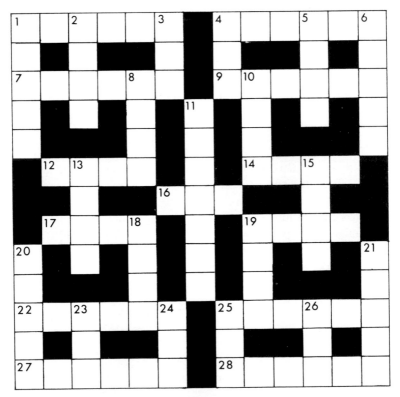

Taken from *New Electronics*, 10 December 1974, with acknowledgements.

2. A couple of solitaire amusements:

(a) First choose a 2-digit number and then choose another single-digit number. You must now try to make your chosen 2-digit number by various operations on the calculator and keying in the single-digit number precisely four times during this process and pressing no other digits and not using a memory. For example suppose you had chosen 43 and 3 then since

$$43 = 33 + \frac{3}{\cdot 3}$$

we can proceed: 3 $\boxed{\div}$ ·3 $\boxed{+}$ 33 $\boxed{=}$

That was an easy one! Now you try some others.

(b) In a similar vein, can you build *all* the whole numbers between 1 and 100 just using 4 threes?

$$1 = \frac{33}{33}$$

$$2 = \frac{3}{3} + \frac{3}{3} \quad \text{or} \quad \frac{3 \times 3 - 3}{3}$$

$$3 = 3^3 \div 3 \div 3$$

$$4 = \frac{3}{\cdot 3} - 3 - 3$$

$$5 = \frac{3 \times 3 + 3!}{3} \quad \text{or} \quad \frac{3^2 - 3}{3} + 3$$

$$6 = 33 - 3^3$$

and so on.

Now what about using 4's or 7's? Alternatively, find how many different ways you can devise to generate each integer rather than being content with a single procedure.

3. This next game is a test of mathematical knowledge. The first player enters a number into the display and passes it to the second player who presses just one function key (e.g. x^2, sin, log) which directly displays a result, and passes the calculator back to the first player who then has to guess which key has been pressed.

4. Nim is a game for two people. In this version the game starts off with, say, 50 entered in the display and each player in turn subtracts a whole number (1 to 7), with the aim of ensuring that it is the other person who first produces zero or a negative number in the display.

Note: There are several books now published on calculator games.

References

Science and Mathematics

Berridge, H. J. J. (1975). The magnetic field of Helmholtz coils. *School Science Review.* Vol. 56, No. 197, pp. 782–6.

Bollen, F. A. (1972). Attitude assessment in science teaching. *School Science Review.* Vol. 54, No. 187, pp. 217–35.

Briggs, S. J. (1971). Hot wheels physics. *School Science Review.* Vol. 53, No. 182, pp. 169–72.

Burghes, D. N. (1974). Optimum staging of multistage rockets. *Intl J. Math. Educ. in Sci. and Tech.* Vol. 5, No. 1, pp 3–10.

Chandler, D. S. (1976). Stepwise approximation to an orbit (revisited). *The Physics Teacher.* Vol. 14, No. 1, pp. 42–4.

Deakin, M. A. B. (1974). An algebraic route to π. *Intl J. Math. Educ. in Sci. and Tech.* Vol. 5, No. 2, pp. 201–7.

Engel, A. (1973). Outline of a problem oriented, computer oriented and applications oriented high school mathematics course. *Intl J. Math. Educ. in Sci. and Tech.* Vol. 4, No. 4, pp. 455–92.

Foglio Para A. and Lazzarini, E. (1974). Some simple classroom experiments and the Monte Carlo method. *J. Chemical Education.* Vol. 51, No. 5, pp. 336–42.

Freeland, P. W. (1976). Ethene and the expansion of immature fruits. *School Science Review.* Vol. 57, No. 200, pp. 493–5.

Haldane, J. B. S. (1953). Some animal life tables. *J. Inst. Actuaries.* Vol. 79, pp. 83–9.

Holdsworth, D. K. (1976). Uses of pocket calculators in chemical education. Part 1, *School Science Review.* Vol. 58, No. 202, pp. 83–9. Part 2, *School Science Review.* Vol. 58, No. 203, pp. 284–7.

Hudson, R. (1965). The spread of the collared dove in Britain and Ireland. *British Birds.* Vol. 58, No. 4, pp. 105–39.

Land, F. W. (1975). *The Language of Mathematics.* (2nd Edition), John Murray.

Lorrimer, J. E., McMullan, J. T. and Walmley, D G. (1976). Searle's apparatus revisited. *Physics Education.* Vol. 11, No. 1, pp. 42–4.

Maynard Smith, J. (1968). *Mathematical Ideas in Biology.* Cambridge University Press.

Mendoza, E. (1975). Pocket calculators and numerical methods in physics teaching. *School Science Review*. Vol. 56, No. 197, pp. 718–33.

Murton, R. K. (1966). A statistical evaluation of the effect of woodpigeon shooting. . . . *The Statistician*. Vol. 16, No. 2, pp. 183–202.

Perry, M. J. (1977). A simulation of first-order chemical kinetics on a programmable calculator. *School Science Review*. Vol. 58, No. 204, pp. 491–3.

Phillips, R. F. (1977). Simple gravitation using a programmable pocket calculator. *Physics Education*. Vol. 12, No. 6, pp. 360–3.

Scott, M. R. (1973). A look at examinations. *Intl J. Math. Educ. in Sci. and Tech*. Vol. 4, No. 2, pp. 97–102.

Siddons, J. C. (1975). More theoretical odds and practical ends. *School Science Review*. Vol. 56, No. 196, pp. 493–506.

Siddons, J. C. (1976). More bits and pieces: a second physics miscellany. *School Science Review*. Vol. 58, No. 202, pp. 12–24.

Smith, J. M. (1975). *Scientific Analysis on the Pocket Calculator*. John Wiley.

Snaddon, R. B. and Runquist, O. (1975). Simulated experiments. *Education in Chemistry*. Vol. 12, No. 3, p. 75.

Turl, E. J. (1976). The magnetic field of Helmholtz coils. *School Science Review*. Vol. 58, No. 202, pp. 106–9.

Wynne Wilson, W. (1976). Using square roots. *Mathematics in School*. Vol. 5, No. 5, pp. 23–4.

Calculators

Cochran, D. S. (1972). Algorithms and accuracy in the HP 35. *H.P. Journal*. June, pp. 10–11.

Cook, J. J., Fichter, G. M. and Whicker, R. E. (1975). Inside the new pocket calculators. *H.P. Journal*. November, pp. 8–11.

Crowley, W. L. and Rode, F. (1973). A pocket-sized answer machine for business and finance. *H.P. Journal*. May, pp. 2–8.

Kelly, S. and MacFarlane, S. (1974). Floating point arithmetic. *Electronic Engineering*. April, May and June.

Liljenwall, E. T. (1972). Packaging the pocket calculator. *H.P. Journal*. June, pp. 12–13.

McWhorter, E. W. (1976). The small electronic calculator. *Scientific American*. Vol. 234, No. 3, pp. 88–98.

Meggitt, J. E. (1962). Pseudo division and pseudo multiplication processes. *I.B.M. Journal*. April, pp. 210–26.

Neff, R. B. and Tillman, L. (1975). Three new pocket calculators. *H.P. Journal*. November, pp. 1–7.

Peterson, K. W. (1974). Testing the HP 65 logic board. *H.P. Journal*. May, pp. 18–20.

Stockwell, R. K. (1974). Programming the personal computer. *H.P. Journal*. May, pp. 8–14.

Taggart, R. B. (1974). Designing a tiny magnetic card reader. *H.P. Journal*. May, pp. 15–17.

Tung, C. C. (1974). A 'personal computer'; a fully programmable pocket calculator. *H.P. Journal*. May, pp. 2–7.

Volder, J. E. (1959). The CORDIC trigonometric computing technique. *I.R.E. Transactions on Electronic Computers*. September, pp. 330–4.

Whiting, T. M., Rode, F. and Tung, C. C. (1972). The powerful pocketful—an electronic calculator challenges the slide rule. *H.P. Journal*. June, pp. 2–9.

Solutions to exercises and Notes

The answers are given to varying degrees of accuracy. Generally speaking more figures are given than is appropriate to quote as a sensible final answer. This is done so that the calculation procedures can be checked accurately and should not be taken as in any way an encouragement to take calculator answers to all the figures displayed, nor to assume that in every case all the figures quoted are correct. Answers which are exactly right are marked as such unless the text makes it clear. Generally the units have been omitted when the context of the problem indicates the appropriate units.

Chapter 2

$$(a+b) \times (c+d) = (a+b) \times c + (a+b) \times d = \left[\frac{(a+b) \times c}{d} + (a+b) \right] \times d$$

so the key strokes are

$$a \boxed{+} b \boxed{\times} c \boxed{\div} d \boxed{+} a \boxed{+} b \boxed{\times} d$$

Chapter 4

4.1 N.B. Some machines carry hidden extra digits which cannot be uncovered

4.2 $32 \cdot 43 \%$ N, $67 \cdot 57 \%$ O

4.3 $1916 \cdot 9$ kPa

4.4 About once an hour

4.5 $4 \cdot 189 \times 10^{-9}$ kg, $7 \cdot 874 \times 10^{12}$

4.11 $2 \cdot 890 \times 10^{-10}$, $4 \cdot 594 \times 10^{-10}$

4.12 $490 \cdot 2$ cm

4.13 $113 \cdot 7$

4.14 $0 \cdot 8593$

4.15 $94 \cdot 49$

4.16 (*a*) (i) $m_1 = 9095$, $m_2 = 905$
      (ii) $m_1 = 7931$, $m_2 = 1703$, $m_3 = 366$
      (iii) $m_1 = 6914$, $m_2 = 2181$, $m_3 = 688$, $m_4 = 217$
      (iv) $m_1 = 6087$, $m_2 = 2419$, $m_3 = 961$, $m_4 = 382$, $m_5 = 152$
      (All figures are rounded to nearest kg)
  (*b*) (i) $6 \cdot 372$ (ii) $7 \cdot 421$ (iii) $7 \cdot 933$ (iv) $8 \cdot 226$

4.17   (a) 441   (b) −299   (c) 330
4.18   The fit is quite good. Plotting '$y = \log$ [population]' against '$x = $ age' would show this
4.19   The fit is moderate
4.20 $P_{20} = 37123 \cdot 5$, 418
4.22   (a) 3, 2, 5, 7, 12, 19, 31, . . .
       (b) 1, 2, 4, 3, 1, 2, 4, 3, . . .
4.25 $t_n \rightarrow \pi$
4.26   6765
4.28   Exponential growth (geometric progression)
4.29 $\cdot 0434782608695652173913$, $\cdot 0384615$,
       $1 \cdot 206896551724137931034482758\dot{6}$, $\cdot 64516129032258\dot{0}$
4.30   The period is often one less than some factor of the denominator. To be more precise, for the fraction p/q where p and q are integers with no common factor then if $q = 2^a \times 5^b \times R$ where R is not a multiple of either 2 or 5 then the decimal equivalent to p/q has m non-recurring digits followed by n recurring digits, where

$$m = \max \ (a, b),$$

n is the smallest integer solution of $10^n = 1$ (mod R)

Chapter 5

5.1   £346.87½
5.2 $430.66
5.3   Not quite: £974.47½
5.4   £443.50
5.5 $1064.54
5.6   (a) simple: £45.00, compound: £53.86; (b) simple: £45.00, compound: £55.13. Simple always the same. Compound always more than simple and most for the 'longer time smaller rate' case.
5.7   4096000
5.8   Yes: £1016.91½
5.9 $6 \cdot 576\%$ p.a.
5.10 £2413.46
5.11 All entries are rounded down rather than rounded off
5.12 £1421.14
5.14 £72.29

5.15 $T = \dfrac{\log \ [12M/(12M - rL)]}{\log \ [1 + r]}$; 7 years

5.18 In one year $100 at $7\frac{1}{2}\%$ p.a. 'instant compounding' becomes $107.79 and at $7\frac{3}{4}\%$ p.a. annual compounding becomes $107.75, so 'instant interest' is better.

In one year \$100 at $6\frac{1}{2}\%$ p.a. 'instant compounding' becomes \$106.72 and with annual compounding of $6\frac{3}{4}\%$ would be 106.75 so in this case annual compounding is better.

5.20 ln (a) means $\log_e (a)$ so $e^{kx} \equiv e^{\ln (a)x} \equiv [e^{\ln (a)}]^x \equiv a^x$

5.23

Year	2	4	6	8	10	12	14	16	18	20
Population	143	187	225	254	273	284	291	295	297	299

Maximum size 300

5.25 $T_{\frac{1}{2}} = 6\cdot129$ hours
5.26 $3\cdot12 \times 10^4$ Pa
5.27 $\lambda_U = 2\cdot94956 \times 10^{-6}$, $\lambda_T = 8\cdot66434 \times 10^{-6}$

Time	0	25	50	75	100	150	200
N_T	0	12·4	21·4	27·9	32·4	37·0	37·8
N_U	193·75	180·0	167·2	155·3	144·3	124·5	107·4

300	400	500	600	700	800	900	1000
33·8	27·6	21·6	16·5	12·5	9·3	7·0	5·2
80·0	59·5	44·3	33·0	24·6	18·3	13·6	10·1

5.28 The answer can in fact be obtained by directly calculating e^{230} which does not cause overflow since $e^{230} \approx 7\cdot722 \times 10^{99}$ (whereas $e^{-230} \approx 1\cdot295 \times 10^{-100}$)

Chapter 6

6.2 8·533
6.4 4·49, 6·10, 14·4
6.5 About 2050 years old
6.6 $3\cdot490 \times 10^5$
6.7 30585
6.8 11.30 a.m.
6.9 11.56 a.m.
6.10 1·956
6.11 (*a*) 1·328 (*b*) 8·509
6.13 (*c*) 151 m
6.14 (*a*) angle correct (*b*) clears by 3 m

6.15 $\theta = 3 \cdot 087702086$; more generally $\theta = \dfrac{180n\pi}{\pi - (-1)^n 180}$

6.16 $33 \cdot 5$

6.18 (a) 3685 n.m.   (b) 9087 n.m.

6.19 $\alpha = 32 \cdot 63°$, $\beta = 44 \cdot 19°$

6.20 $i = 59 \cdot 38°$, $r = 40 \cdot 19°$, refractive index $= 1 \cdot 334$

6.22 (a) $(32 \cdot 82, 64 \cdot 42)$ (b) $(-3 \cdot 61, -2 \cdot 27)$

6.23 (a) $(7 \cdot 78, -110 \cdot 3°)$ (b) $(6 \cdot 31, -3 \cdot 63°)$

6.25 $326 \cdot 7$ knots on a bearing $304 \cdot 5°$

6·26 $2 \cdot 95$ N acting at an angle $169 \cdot 5°$

6.28 $50 \cdot 52$

6.30 $28 \cdot 75754$, 0, $7 \cdot 573239$

Chapter 7

7.1 $207 \cdot 21$

7.2 $20 \cdot 125$ (exact)

7.3 $16 \cdot 81984$

7.4 (Geometric Mean) $25 \cdot 8°C$, $30 \cdot 9°C$

7.5 (Harmonic Mean) 491

7.7 125 (exact), $17 \cdot 53846$, $4 \cdot 187895$

7.8 $0 \cdot 8365432$

7.9 $25 \cdot 89$ (exact), $0 \cdot 6788962$; $25 \cdot 89$ (exact), $0 \cdot 7156194$;
$25 \cdot 4374 \leqslant \mu \leqslant 26 \cdot 3426$; $25 \cdot 2111 \leqslant \mu \leqslant 26 \cdot 5689$

7.10 $14 \cdot 45$, $13 \cdot 53969 \leqslant \mu \leqslant 15 \cdot 36031$

7.11 slope $0 \cdot 5029$, intercept $33 \cdot 84$, $\therefore$ $y = 0 \cdot 5029x + 33 \cdot 84$

7.12 $\rho = 1$ (exact), $r = 0 \cdot 9652$

7.13 Paper 1: mean $64 \cdot 47$, s.d. $21 \cdot 79$; Paper 2: mean $56 \cdot 32$, s.d. $17 \cdot 60$;
$r = 0 \cdot 7997$

7.14 $\chi^2 = 6 \cdot 00775$, not significant at 5% level

7.15 (a) $\chi^2 = 9 \cdot 7232$, significant at 5% level
(b) $\chi^2 = 3 \cdot 3352$, not significant at 5% level

7.16 $\chi^2 = 61 \cdot 42$, highly significant

Chapter 8

8.1 21 days

8.2 5

8.3 (a) 56   (b) 200

8.4 (a) $0 \cdot 248$ (b) $0 \cdot 231$

8.5 23 People, 57 People

8.6 (a) (i) $95 \cdot 399$ (ii) $94 \cdot 902$ (b) 225

8.7 (a) 1, 10, 45, 120, 210, 252, 210, 120, 45, 10, 1

8.8 $0 \cdot 1122752$ (≈ 1 in 9)

8.9 (a) $0 \cdot 2016$ (b) $0 \cdot 3232$ (c) $0 \cdot 2587$ (d) $0 \cdot 2165$

8.10 $0 \cdot 28$ (≈ 2 in 7)

8.11

r	0	1	2	3	4	5	6
p	0·133	0·271	0·273	0·182	0·090	0·035	0·011

8.12   0·2019, 0·3230, 0·2584, 0·2167

Chapter 9

Plotting graphs example:

x	0·0	0·5	1·0	1·5	2·0
$x^2 e^{-x} + \sin x$	0·00	0·63	1·21	1·50	1·45

x	2·5	3·0	3·5	4·0	4·5	5·0
$x\, e^{-x} + \sin x$	1·11	0·59	0·02	−0·46	−0·75	−0·79

9.3   Maximum occurs when $R = r$
9.4 $r = \frac{2}{3}$
9.6   (b) Velocities are increased—curve is pushed to the right
9.9   (a) Continuous   (b) continuous   (c) continuous
       (d) continuous at $x = 5$, not at $x = 1$ (e) continuous
       (f) not continuous but limit exists
       (g) no left or right limits so not continuous   (h) continuous
       (i) not continuous but limit exists   (j) continuous
       (k) continuous
       (l) left and right limits different so not continuous

Chapter 10

10.1   (a) 0·88332094, 0·24510733
       (b) 7927, 0·0002 (both exact)
10.2   31·55%
10.3   −2·52891796, 0·167449191, 2·36146877
10.4   −3·58913269
10.5   (i) $5·69510 \times 10^{-4}$, $3·96876 \times 10^{-4}$, $3·214 \times 10^{-6}$
       (ii) $7·86684 \times 10^{-4}$
10.6   converges to 0·5
10.7   will not converge to 2·5
10.9   0·46426688
10.12 1 and 2·5 (both exact)
10.13 2·166268

10·14   (a) 1·188914, 2·629850, −0·8187637 (only first physically possible)

(b) −4·415214 (one real root, not physically possible—sphere sinks)

10.15 $x = 1·001000$, $p = 1001$ which is impossible *or* $x = 0·2028461$, $p = 2028$ giving 20·26%

10.16   (a) $x = n\pi$ for n any integer except zero

(b) $x = 0$

10.17   0·5929621 radians

10.18   5·2065755

10.19   2·365020

Chapter 11

11.1   596·0884 g bread, 101·0623 g cheese, 539·4265 g tomatoes

11.2 $\phi_1 = 1·718617 \times 10^{-3}$, $\phi_2 = -8·679657 \times 10^{-4}$
$\phi_3 = 1·313832 \times 10^{-3}$, $\phi_4 = -5·182161 \times 10^{-4}$

11.3 $a = 500$, $b = 516\frac{2}{3}$, $c = 483\frac{1}{3}$, $d = 166\frac{2}{3}$, $e + f = 833\frac{1}{3}$ (all exact)
Actually the fourth equation, involving d, can be omitted.

11.4 $\omega_e = 2990·95$, $\omega_e x = 52·8186$, $\omega_e y = 0·224366$, $\omega_e z = -0·012175$

Chapter 12

12.1   The population decreases (exponential decay)

12.2   166 or 167 ($166\frac{2}{3}$ theoretically)

12.3   (a) 650, 423, 275, 179, 116, 75, 49, 32, 21, 14, 9, 6, 4, 3, 2, 1

(b) 650, 639, 634, 633, 632, 632, 632, . . .

12.4   (a) 53200, 120317, 175205, 139733, 170922, 145102, 168141, 148352 . . . tending to $158333\frac{1}{3}$

(b) The ladybirds die out

12.5   (b) Population alternates between about 128 and 238

12.6   (a) about 780   (b) about 500

12.7   (a) converges rapidly to 267

(b) small decreasing oscillations about 267

(c) large uniform oscillations about 267

(d) small decreasing oscillations about 267

12.8   Limit is 45

Chapter 13

13.1   0·21128865 (analytic: 0·21128856)

13.2   analytic: 1576948·1

13.3   −0·68565138 (analytic: −0·68565128)

13.4   analytic: −8623·1887

13.5   analytic: $2·8934426 \times 10^{10}$

13.6   (i) 9·8385716   9·2056363   8·9116276

(ii) 8·7359390   8·6574428   8·6379196

(iii) 8·6290973   8·6312774   8·6314118

13.10 $-1\cdot282859$

13.11   Trap: $0\cdot52592755$ Midpt: $0\cdot46579823$

13.12   Trap: $0\cdot44711216$ Midpt: $0\cdot46043703$

13.13   (a) $0\cdot48269902$ (b) $0\cdot47973785$

13.14   2: $0\cdot48121843$ 4: $0\cdot48083248$
        8: $0\cdot48073479$ 16: $0\cdot48071028$

13.15   4: $0\cdot31828471$ 8: $0\cdot31555964$

13.16 $1\cdot6583476$

13.18 $0\cdot4759$

13.19 $2\cdot06259$

13.20   Boole: $0\cdot5033$ Six Point: $0\cdot5479$ (Correct value $0\cdot6499$)

13.22   Trap: $1\cdot468$ Simpson: $1\cdot449$

13.23 $2\cdot2$ to $3\cdot8$: $1\cdot16$; $5\cdot2$ to $7\cdot4$: $1\cdot33$

Chapter 14

14.1   (b) $V_x = y$, $V_y = -x$; circular orbit
       (c) (i) Period $4\cdot84$, semi-major axis $0\cdot84$
           (ii) Period $6\cdot28$ (2π), semi-major axis $1\cdot00$

14.3   The inverted equation $a^3/\tau^2 =$ constant may be easier for computation

14.4 $31\cdot9$ ms^{-1}

14.5   (a) Analytic: $10\cdot0986$
       (b) For $dv/dt = g - \beta v - \alpha v^2$, $v = 0$ at $t = 0$, the solution is:
           $v = (k/2\alpha) \times [(e^{kt} - 1)/(e^{kt} + 1)]$
           where $k = \surd(\beta^2 + 4\alpha g)$
           and t is time
           The terminal velocity is therefore $k/2\alpha$

14.9   (a) It is reduced to compound interest with compounding every $0\cdot5$ time units

Solution to Crossword

To read the solution:

   (i)   Perform calculation from left to right as it appears without regard to algebraic hierarchy. Powers apply just to the associated number. 'Sum of products' logic machines will not give the desired results unless intermediate ☐ key depressions are made.

  (ii)   Invert calculator.

 (iii)   Read solution.

ACROSS

1 $131 \times 30 + 1 \times 5^3$

4 $23 \times 39 - 4 \times 31 - 4 \times 20$

7 $2 \times 10^4 - 1003 \times 20 - 21$

9 $4^3 - 1 \times 44 + 1 \times 11^2 + 4$

12 $3 \times 7 \times 11 \times 13 + 77$

14 $3 \times 5 \times 7 \times 29$

16 $113 \times 3 - 1$

17 $77 \times 101 - 59$

19 123×5^2

22 $5 \times 7 \times 11 \times 13 \times 107 - 3^3$

25 $2^2 \times 3^2 \times 61 \times 173$

27 $11^3 - 260 \times 500 + 7$

28 $2 \times 19 \times 41 - 1 \times 4 \times 61 + 1$

DOWN

1 $11 \times 37 \times 9 - 2 \times 15$
2 $109 \times 11 \times 5 - 2$
3 19×16
4 $22 \times 113 - 1 \div 7 \times 2$
5 $7 \times 10 \times 17 + 1 \times 3$
6 $103 \times 150 + 19 \times 5$
8 $6 \times 31 - 13 \times 22 + 1$
10 $151 \times 19 - 5 \times 2 + 5$
11 $159^3 - 47^3 - 13^3 - 11^2$

13 $7^2 + 12^3 \times 2^2$
15 $59 \times (5!)$
18 $7 \times 3 \times 167$
19 $9^3 - 3^3 + 2^4 - 3 \div 10^3$
20 $272^2 - 13^2 + 2$
21 $12^3 + 3^3 + 2^3 \times 20 + 120$
23 $13^3 - 11^3 - 7^3 - 2^3$
24 $6^4 - 5^4 - 4^4 + 3^4 + 3^2$
25 $123 \times 8 - 46$
26 $12^3 - 9^3 - 10^2 + 5^2 - 2^2 - 1$

Index

THE WYKEHAM SCIENCE SERIES

†(*Paper and Cloth Editions available.*)

THE WYKEHAM ENGINEERING AND TECHNOLOGY SERIES

All orders and requests for inspection copies should be sent to the appropriate agents. A list of agents and their territories is given on the verso of the title page of this book.